*f*P

ALSO BY BRUCE BERKOWITZ

American Security (1986)

Calculated Risks (1987)

WITH ALLAN GOODMAN:

Strategic Intelligence (1989)

The Need to Know: Covert Action and American Democracy (1992)

Best Truth: Intelligence in the Information Age (2000)

THE NEW
FACE OF WAR

How War Will Be Fought
in the 21st Century

Bruce Berkowitz

The Free Press

NEW YORK · LONDON · TORONTO
SYDNEY · SINGAPORE

THE FREE PRESS
A Division of Simon & Schuster Inc.
1230 Avenue of the Americas
New York, NY 10020

For information about special discounts for bulk purchases,
please contact Simon & Schuster Special Sales:
1-800-456-6798 or business@simonandschuster.com

Manufactured in the United States of America

1   3   5   7   9   10   8   6   4   2

Library of Congress Cataloging-in-Publication Data

ISBN 0-7432-1249-5

Dedicated to the memory of
Thomas P. Rona,
Frank B. Horton III, and
John R. Boyd

Frank and explicit—that is the right line to take when you wish to conceal your own mind and confuse the minds of others.

—BENJAMIN DISRAELI

Confusion to the enemy!

—ED BALL

CONTENTS

PREFACE

This book is intended to give readers a better understanding of the new face of war and what the United States must do to prepare for it. The main message is this: the Information Revolution has fundamentally changed the nature of combat. To win wars today, you must first win the information war.

Recent experience bears this out. All of the military successes of the United States during the past twenty years have had a common ingredient: the ability to achieve an information advantage over our adversary. Every military failure has occurred mainly because we failed to secure that advantage.

The ability to know where your opponent is and act before he can act has become the most important factor in modern warfare. The ability to hide—and mass to kill when required—has become key to survival.

The ability to use U.S. military forces in military action in the Persian Gulf, deter aggression in the Far East, prevent proliferation in Southwest Asia, or counter global terrorism all depend on a common question: Have we established the information advantage we need for victory?

For the most part, the Information Revolution has treated the United States kindly. The American military is overwhelmingly stronger than any of its potential adversaries. This is mainly because it has been more successful in taking advantage of information technology, and because of American leadership in most information industries.

However, experience has also demonstrated that this margin, while large, is fragile. Much of the technology that was critical to our information edge just ten years ago is now available to all. Also, success in the information wars often depends as much on creativity and ingenuity as on technology.

Traditional concepts of armed self-defense are sorely tested when armies must hide and disperse to survive, and striking first is all-important. Democratic oversight is inherently harder when victory depends more than ever on stealth and secrecy. It will be a challenge to maintain both our security and our values. Understanding the issues, and what is at stake, is the first step in meeting these challenges.

THE NEW
FACE OF WAR

Chapter 1

THE NEW TERRAIN

The next wars will be fought not just on battlefields but also in the world's computers and communications systems. The combatants will often be familiar military powers—like China, France, and Russia—but there will be others, including underestimated military powers, like India; presumed allies, like Israel; and countries that hardly seem to have any military capability at all, like the Philippines. Terrorist groups, landless peoples, and international criminal organizations will also be players in the new warfare. They are laying the groundwork for this war as you read these words. And so are we.

This book is about the future of battle. It is about the threats the United States will face in the years ahead, and how we must prepare for them. It focuses especially on how the Information Revolution is defining these threats and providing the solutions for dealing with them.

To be sure, information technology has been with us for around 4,000 years, since Sumerians started preserving their beer recipes on clay cuneiform tablets. Even electronics has been around since 1844, when Samuel Morse tapped out, "What hath God wrought?" in the first telegram, sent from Washington to Baltimore. There is always a lot of evolution behind most revolutions when you look at them closely.

Yet clearly something new is happening today. Information technology is improving at an exponential pace, and is penetrating every corner of the world. It may be the mainframe that your bank uses to balance its books. It may be an old PC that someone has trucked in through a mountain pass in the back of a Datsun pickup, or a microprocessor built into the control system of an electric generating plant. Or it may be a disk with pirated software sold on the streets of Hong Kong. But it is there, in every sector of society—finance, transportation, utilities, entertainment.

Not only is information technology everywhere; it continues to grow more and more powerful and, in the process, takes new forms. Booking a flight today is almost always easier on the Internet—only the most visible information technology of the last decade. Even if you call the airline, you will speak first to a machine, a product of cheap microprocessors and even cheaper read-only memory chips. When you do reach a real person, he or she will be in another state or even another country, a result of fiberoptic digital communications, which allow companies to locate their operations wherever they can find the best labor and real estate deals.

No one has felt the effects of the Information Revolution more than the world's military forces. This is no surprise, considering how much of the technology was developed in military labs or under defense contracts. The money that created microchips, satellite navigation, and the Internet was not green; it was olive drab, sky blue or navy blue—or deep black. The government was paying the bills in the name of national security.

Information technology has become so important in defining military power that it overwhelms almost everything else. Back in the Cold War, everyone knew technology was a "force multiplier" (to use the jargon of the day), but no one knew by how much. Did better technology make a U.S. battalion half again as effective as its Soviet counterpart? Twice as effective? Recent experience suggests that the right technology, used intelligently, makes sheer numbers irrelevant.

The tipping point was the Gulf War in 1991. Iraq had the third-largest army in the world and had just killed 300,000 Iranians in eight years of the most gruesome warfare of the twentieth century. American generals faced the same problem as their Iranian counterparts: how to dislodge dug-in Iraqis from well-prepared, fortified positions. Officials feared catastrophe and carnage. But when the war was over, the United States and its Coalition partners had lost just 240 people.

Iraq probably suffered about 10,000 battle deaths, although no one will ever really be sure. The difference was that the Americans could see at night, drive through the featureless desert without getting lost, and put a single smart bomb on a target with a 90 percent probability. The difference was information technology.

The same thing happened when the United States fought Yugoslavia in 1999 and the Taliban regime in Afghanistan in 2001. Each time experts feared the worst; each time U.S. forces won a lopsided victory (at least on the battlefield; the strategic objectives remained more elusive).

There is a trend. Information technology is so important in war today that it overwhelms everything else. And when it comes to applying information technology to warfare, today no one is better than the United States.

But will the United States always enjoy "the Information Edge"? In the Gulf War the Americans were the only ones who could see at night and navigate the desert. Today anyone can purchase a night-vision scope and satellite navigation gear by mail order. Anyone with a credit card and a reasonably fast Internet connection can download satellite imagery from the Web. The whole package costs less than $10,000.[1] Our government may try to control this technology, but experience suggests it will usually fail. And simply *having* better technology does not guarantee success. Victory goes to the side that understands how to *use* information technology more effectively.

The most significant effect of information technology on warfare has been to make the concept of "the front" obsolete. Everyone and everything is part of the battlefield today and a potential target. True, we have heard this before. In the 1930s military pundits wrote about future aerial gas attacks against cities. In the Cold War, children practiced their "duck and cover" drills to prepare for a nuclear strike. There was always some new weapon (like the bomber) or tactic (like blitzkrieg) or concept (like guerrilla war) that would put everyone on the front line.

This time is different. The front line really is disappearing from war. It is not just that the battlefield is becoming bigger because weapons have longer range and are more powerful (which they are). The difference today is that worldwide communications enable armies to disperse—and even deploy covertly within their adversary's territory before a battle even begins.

At the same time, weapons are so accurate—another result of information technology—that armies *must* disperse. Today, if you can see a target, you can usually kill it. Often you do not even need to see it; you simply need to know where it is, or where it will be at some moment in the future. In the past armies massed for mutual protection. Now any army that masses offers an easy target. So armies must hide. Concealment and dispersion become their normal operating status, and, if the profile of military forces is lower, the profile of civilians and the surrounding environment become higher, and thus the front is gone.

Of course, some things about war will never change. Wars will always be bloody. Soldiers—and, inevitably, civilians—will be burned alive, torn to pieces, and often die horrible deaths. Combat will always be confusing and terrifying. Even in the Information Age, war will often come down to a face-to-face, hand-to-hand encounter where one soldier prays he can snuff out the life of his enemy and save his own. High-tech warfare is not antiseptic warfare. Look at Afghanistan. Look at the West Bank.

Nevertheless, information technology is now the essential difference between winning and losing. Better technology means deadlier armies. And as armies become more dependent on information technology, they will develop new kinds of vulnerabilities.

In considering the new face of war, it is useful to keep five numbers in mind. They draw the basic outline of American national security today.

The first number is $750 billion. According to the CIA's *World Factbook,* that's the total of the world's military spending—that is, every country's defense budget combined. This number has been declining, slowly, almost every year since the end of the Cold War, partly because Russia can no longer pay for its military, and partly because everyone else no longer fears the Russians.

The second number is $380 billion. That's approximately what the United States spends annually for its military forces today. In other words, the United States now spends roughly as much on defense as everyone else in the world combined.

Rarely in history has a single country been so dominant. During the Cold War, pundits argued whether the Soviet Union or the United States was ahead in the so-called arms race. Today there isn't even a race.

To be sure, there is lots of room to cut waste and improve efficiency. Many programs do nothing more than represent bureaucratic turf and congressional pork. But partly because it spends this much money, the United States can do things no one else can, such as build aircraft invisible to radar, design bombs that can hit within a few feet of their target, and transport thousands of troops halfway around the world.

The third number is 3.2 percent. That is the approximate percentage of the U.S. gross domestic product that today goes to defense. Think of this as a measure of the "defense burden," or how much of our wealth that might be spent on more productive activities is instead going to fund the military. The most interesting thing about this statistic is that it suggests that our defense burden today is remarkably small by historical standards. During the Cold War we spent up to four times as much of the nation's wealth annually on defense.

Put these three numbers together, and the picture looks something like this: Not only does the United States have overwhelming military power; we are keeping our edge without even breathing hard. We could keep up our current defense effort—and even a bit more—indefinitely, and the U.S. economy would hardly notice the difference.

However, if 3.2 percent of the U.S. economy is devoted to defense, it follows that 96.8 percent is *not*. This is another way of quantifying the obvious: The Defense Department is a small part of American life. Most Americans have little direct involvement in national defense. They know little about their armed forces. For that matter, the typical American is blithely unaware of the threats that face us. The typical college student—no, the typical college professor—could not tell you the difference between Ahmed Ressam and Ahmed Barada. (The former is a convicted Al Qaeda terrorist from Algeria who planned to bomb Los Angeles International Airport. The latter is a professional squash player from Egypt.)

This is a problem, because today we need cooperation between the government and the private sector more than ever before—both for ensuring access to the information systems that our foreign adversaries use, and for protecting our own. But cooperation is a two-way street.

In Information Age wars, victory usually goes to the side having more influence over technology and better access to the world's electronic infrastructure. And *that* often depends on market share. If U.S. companies do not dominate the nets, someone else will, and that would be to our disadvantage. When legislators pass laws and civil servants implement regulations, today they all need to think about how they are shaping the electronic battlefield.

Of course, it would also help if the men and women heading American companies had a better understanding of the pivotal role they play today in the nation's security. Being business executives, their understanding will be highly correlated with economic carrots and sticks. And, if warfare in the Information Age is often going to spill over into the private sector, we all need to think about how to ensure effective oversight and protect civil liberties, so that we do not destroy democracy in the process of defending it.

The fourth number to keep in mind is 17 percent. That's the annual rate at which China's defense budget has been growing lately. It reminds us that, despite America's predominance, there are countries willing and able to make big investments to challenge us.

Oddly enough, starting from behind has certain advantages. When you design a new army from scratch, you start with a blank slate. You can learn from your competitors and avoid their mistakes. Though smaller, unit for unit your army will be newer. Also, an economist would say that most of the huge U.S. military establishment is a "fixed cost." Politicians and military officials are both reluctant to shut down production lines, retire existing weapons, and fire people. So most of the $380 billion dollars the United States spends annually on defense does not vary much. Changes occur only in small increments each year. That makes it harder for the U.S. military to adapt.

Our adversaries can start afresh. They can focus on their specific needs and our greatest vulnerabilities. So U.S. military superiority may be less certain than simple dollar comparisons suggest. And because military power depends so much today on information technology, any advantage the United States enjoys today could disappear quickly.

Anyone who has shopped for a computer knows how quickly information technology becomes cheaper and more capable. As military power becomes more closely linked to information technology, any investment becomes outdated faster. Just as it's hard to keep up with the neighbors in having the fastest computer on the block, it's hard to keep up with the neighbors in having the most capable Information Age military force.

Besides, our adversaries know they cannot match the United States in tanks, planes, and warships. They know they will most likely lose any war with us if they play according to the traditional rules. So, naturally, they will try to change the rules.

This is why most military dangers we face from abroad today are from "asymmetric threats"—strategies and tactics that avoid our strengths head-on, and instead hit us where we are weak. Our adversaries will use unconventional weapons and tactics to overcome our advantage.

For example, consider the potential threat of a Chinese attack on Taiwan. Chinese leaders routinely claim Taiwan is Chinese territory ruled by a "renegade regime." They openly say they plan to recover it, even if this requires war. President George W. Bush, on the other hand, has said that the United States will do "whatever it takes" to prevent China from conquering Taiwan. The loss of a popularly elected democratic government to an authoritarian regime would be an epochal event that we could not let stand.

U.S. armed forces are much larger and better equipped, but China enjoys several advantages. U.S. forces must travel 8,000 miles to defend Taiwan; Chinese forces only need to travel 100 miles to attack it. Most Chinese people, regardless of what they think of the Communist regime, think reuniting Taiwan with the mainland is important. The typical American has only a vague idea of where Taiwan is located.

China would likely strike American forces where they are weakest. They would probably attack the bases on the Pacific Rim that are essential to any U.S. military operation in the region. Chinese leaders also understand our dependence on information technology. If the Chinese military can neutralize certain essential U.S. computers and communications links, then U.S. plans to defend Taiwan might collapse like a house of cards.

The fifth and final number is 3,025—the number of people killed at the World Trade Center, the Pentagon, and Somerset County, Pennsylvania, on September 11, 2001. This number reminds us that although today the United States is the supreme military power, hostile countries and terrorist groups can still strike directly at America from halfway around the world with devastating results.

The September 11 attacks showed how some basic features of war have changed. Small countries and even organizations that are not states can successfully strike great powers. Also, these threats can conceal themselves and their plans for maximum advantage.

To deal with these threats, the United States must not only beat them in the information war—U.S. leaders must also be able to decide when and how to strike them before they strike us. Doing this while observing the traditional rules of war will be a challenge. So will maintaining democratic control of U.S. armed forces. Welcome to the realities of warfare in the Information Age.

Chapter 2

AN AFGHAN HARD DRIVE

Alan Cullison realized he needed a new computer. Ordinarily this would not have been a problem, except that at the time Cullison happened to be in the Hindu Kush mountains in eastern Afghanistan. It was October 2001, his truck had just fallen off the road, it was cold as hell, and he was still forty miles from Kabul.

Cullison had been in Afghanistan covering the war against the Taliban as a reporter for *The Wall Street Journal*. "I was coming over the mountain and our truck lost its brakes. We hit another truck, rolled over, and my computer was smashed." He spent the next month reading his reports to New York over a satellite phone.

In November the Taliban's resistance collapsed, and Cullison hitched a ride to Kabul. He hoped to find a replacement for his laptop. "There aren't a whole lot of computers in Afghanistan," Cullison recalls. But with the help of a local shopkeeper he found someone trying to sell a Compaq laptop and an IBM desktop. Apparently the current owner had acquired the two machines from the local headquarters of Mohammed Ataf, chief strategist of Al Qaeda, the international terror network. By bankrolling the Taliban, Al Qaeda had virtually taken over the country, and had started the war back in September with its attacks on New York City and Washington.

Ataf no longer needed the computers, having been killed in a U.S. bombing raid. The Taliban started a rapid retreat back to Kandahar once the Northern Alliance defeated them first at Mazar-e-Sharif and again at Kunduz. In fact, the war was turning into a rout as the Northern Alliance started to roll toward Kabul. The computers got left behind. A looter fenced them to a local trader, who was now offering them for sale to Dow Jones, Inc., for just $4,000.

Cullison called John Bussey, his editor in New York. Bussey said the price seemed high and told him to get a better offer. Cullison haggled. He got the two computers for $1,100.

In fact, Cullison knew that the computers were worth more than their face value. While he was shopping for a computer, he learned from one merchant that some of the local Arabs—most likely, members of Al Qaeda—had brought their computers to him for repair. The Arabs always looked over his shoulder when he worked on their machines, and their files were usually password-protected.

Cullison put two and two together. He was already behind the other reporters who had beat him to Kabul and grabbed every piece of paper they could find in Al Qaeda's abandoned offices. Making a virtue of necessity, he decided that, as long as he needed a computer, he might as well go hunting Al Qaeda hard drives.

That was how he came upon the IBM desktop. Cullison found what he was looking for. The computer held four years' worth of records detailing operations of the terrorist organization in Afghanistan and abroad.[1]

If you believed some experts after the September 11 attack, Al Qaeda was hard to penetrate because it was a loose, amorphous network of groups that used only person-to-person contacts. It was hard to track because the organization had sworn off technology. It sounded like a good explanation and played into the popular bigoted image of Al Qaeda's consisting of a bunch of rag-heads riding camels through mountain passes, carrying secret instructions written in invisible ink on goat scrotum.

Hardly. Al Qaeda was a modern army. It was as adept with computers as any organization founded by the engineer son of a construction millionaire and staffed largely by middle-class educated males. Intercepting Al Qaeda communications was hard mainly because the organization understood information technology so well. It had some of the best operational security anywhere—OPSEC, to use the technical term.

Al Qaeda had tightened its security even before the press reported that U.S. intelligence listened in on Osama bin Laden's satellite telephone calls to his mother. Al Qaeda members in Europe and North America began to e-mail from cyber cafés and libraries. They encrypted their files and would use a cell phone for a week or two, throwing it away before they could be traced. They knew that the flow of digital data traveling around the globe had grown exponentially during the past decade, and they knew how to hide in the torrent.[2]

Meanwhile, Al Qaeda operatives used the Internet to scope out targets. They downloaded layouts of bridges and buildings from Web sites. In the past, collecting this kind of information might require traveling around the world. Getting it to someone in the field required undercover couriers. Now you could click, get the data, click again, and send the diagrams to a temporary, untraceable e-mail address.

Anyone who had bothered to follow a trial in the Southern District of New York that spring might have predicted that such a computer existed somewhere in Kabul. In August 1997, three years earlier, Special Agent Daniel Coleman, some FBI colleagues, and several Kenyan police officers had dropped in on a white stucco house at 1523 Fedha Estates, in Nairobi. It was the home of Wadih el-Hage, an assistant to bin Laden. It was also the office el-Hage used for Al Qaeda's African operations.

El-Hage, a Lebanese Catholic, grew up in Kuwait, converted to fundamentalist Islam, and moved to the United States to attend college. He left in the early 1980s to fight the Soviets in Afghanistan with the mujahideen. That was where he hooked up with Osama bin Laden. Returning to the United States, he became a citizen and, at about the same time, got involved with Islamic fundamentalist groups.[3] By the early 1990s el-Hage was in Africa, working for bin Laden—first in Khartoum, Sudan, and then in Nairobi.

A Sudanese source tipped off U.S. officials that el-Hage had been involved in the 1993 World Trade Center bombing, and so the FBI got a warrant to search his home in Nairobi. El-Hage was not in, but the FBI agents did find his Apple PowerBook 140. Coleman gave the computer to Special Agent Robert Crisalli, a computer technician who had joined the FBI after a stint at Oppenheimer Capital. Crisalli hooked up a Zip drive to the laptop and booted the computer using his own operating system. Then, using a software toolkit on the Zip drive, he downloaded a mirror image of the computer's hard drive to a portable optical disk.

Once Crisalli captured the contents of the computer, he went back to Washington and burned the files onto CDs—sort of a "Best of bin Laden"—that could be sent to FBI analysts anywhere. In May 2001 one of the CDs become Exhibit 300T in *United States of America v. Usama bin Laden, et al.*, the trial of el-Hage and three other Al Qaeda suspects accused of planning the 1998 bombings of U.S. embassies in Nairobi and Dar es Salaam.

Even after the September 11 attack some people still could not believe Al Qaeda was a real army. "How do you penetrate an organization which is largely ideological and bound by religious fervor?" wondered Larry Johnson just after the attack. Johnson was an apt person to ask the question, having served as Deputy Director of the State Department's Office of Counterterrorism during the first Bush administration. "They don't have a membership. You don't have to fill out an application. It's not like joining a country club."[4]

That would have been news to Jamal Ahmed al-Fadl, who had recently turned state's evidence and testified against el-Hage. Prosecutor Patrick Fitzgerald asked al-Fadl how he came to join Al Qaeda in 1990. "I swear and I signed," he said. "He give me three paper, I read it, and after that I swear in front of him and I sign the papers," al-Fadl added, noting that he was just the third person to sign a contract adding him to the organization. Apparently in Al Qaeda there is cachet in having a low membership number. Just like joining a country club.[5]

El-Hage denied involvement with Al Qaeda, but letters on the computer told a different story. Several described the preparations for the bombing and called el-Hage the "engineer" of the Nairobi cell. The jury convicted el-Hage of conspiracy and perjury. The judge sentenced him to prison for life.

Even before these computers turned up in Kabul and Nairobi, yet another computer provided the tip-off for an earlier Al Qaeda operation, an operation that had several parallels to the September 11 attacks.

In January 1995 Aida Fariscal was the watch commander at Police Station No. 9 in Manila. The fire department reported that a bunch of Pakistanis playing with firecrackers had set off a fire in their apartment down the street. The Manila police were already tense. Pope John Paul II was scheduled to visit the Philippines soon. Everyone was concerned about a possible terrorist attack; John Paul, after all, had already been the victim of an assassination attempt back in 1981.[6]

Fariscal decided to investigate. When she and her officers entered the apartment, they discovered what appeared to be a makeshift lab for concocting explosives. The two men in the apartment, Ramzi Ahmed Yousef and Abdul Hakim Murad, tried to explain why they were storing hot plates, electrical wire, and large plastic containers filled with chemicals in their bachelor pad.

The apartment turned out to be a base for an Al Qaeda cell. This was before most people had heard much about the terrorist network, so at first Fariscal and the police officers did not know what to make of the scene. They also did not know that Yousef was wanted for planning the 1993 World Trade Center bombings and carried a $2 million bounty on his head.

Murad and Yousef tried to make a break for it, and in the confusion that followed, Yousef escaped. But Murad tripped, and the police tackled him, tied him up, and hauled him to the precinct house. Before leaving the apartment, Fariscal also found a laptop computer, which later turned out to belong to Yousef. Suspecting that something bigger was going on, the Manila police called in the Philippine national authorities, and they called in the local FBI and CIA representatives.

Although the computer's hard drive was password-protected, the Philippine authorities were able to open at least some of the files and translate them from Arabic. They contained plans for Operation Bojinka, a plot to plant bombs on a dozen airliners en route between the United States and Asia. The bombs would go off at approximately the same time, while the aircraft were over the ocean.

Most of the information needed to put the picture together was on that magnetic disk: true names, aliases, and telephone directories. The files also revealed that Murad had been taking flying lessons and that the organization was planning a suicide strike on Washington, possibly CIA headquarters. The FBI had enough information to track down Yousef in Pakistan; he was later extradited to the United States, tried, and convicted for his connection to the 1993 World Trade Center attack.

Like the Nairobi and Manila laptops, the Kabul computers contained valuable information. Many of the files were encrypted, but with some help, Cullison was able to extract about 1,800 files. Most of them were routine memos, records, and letters. This is typical in intelligence analysis. Every once in a while a single piece of evidence captures the entire story—say, the master plan of an attack. But usually you just get cryptic bits and pieces: "Bob wants to meet Jake at the diner." It takes some puzzle solving to figure out Bob and Jake are pseudonyms for two terrorist operatives, and the diner is a safe house.

Pieced together, the odds and ends from Cullison's computer provided a mosaic of Al Qaeda at work over a four-year period: lists of members, locations of cells, methods for moving people and money. One of the files was a scouting report from August 2001 sent by an "Abdul Ra'uff." It described several public buildings in Israel—targets. You needed to read through the jargon and pseudonyms, but it was clear that Abdul Ra'uff was an alias for Richard Reid, a British citizen of partly Jamaican descent who had recently been arrested in an incident that was, at least by pre–September 11 standards, unusual.

Reid had been a passenger a few weeks earlier on American Airlines Flight 63 from Paris to Miami. Hermis Moutardier, a flight attendant, smelled something burning. Thinking someone might be violating the no-smoking rule, she looked for the source. A passenger pointed to a scruffy, longhaired passenger—Reid—who quickly put a just-extinguished match into his mouth. Moutardier called the captain.

Reid took off a shoe and lit another match. Moutardier saw a fuse hanging from the shoe and tried to grab it. Reid shoved her into a bulkhead. Moutardier ran to get some water and yelled for help. Cristina Jones, another flight attendant, heard Moutardier and also tried to get the shoe away from Reid; he bit her. By that time other passengers joined in, wrestled Reid to the floor, and tied him up. A doctor injected him with a sedative, and the police arrested Reid when the plane made an emergency landing at Logan International Airport in Boston.[7]

Late-night comics joked about Reid, the "shoe guy" who tried to light up his high-tops on a no-smoking flight. But if Moutardier and Jones had not tackled Reid (six feet four, two hundred and something pounds), Flight 63 would have been lost midway over the Atlantic without a trace. Investigators might not have any idea of what had happened for months. Following on the heels of September 11, such a disaster could have shut down air travel for weeks.

It was proof that Al Qaeda was still several steps ahead of U.S. intelligence. Several days after Reid was arrested, government authorities still believed that he was a lone ranger, possibly a copycat. Only later did they discover, partly with the help of Cullison's computer, that Reid was an alumnus of Al Qaeda's training camps in Afghanistan.[8]

The data was out there. In effect, in late 2001 Al Qaeda and the United States were in a race. The question was whether U.S. intelligence could figure out where the data was, collect it, and analyze the information before Al Qaeda could carry out its attack plan. Al Qaeda won the race.

It is impossible to talk about American national security in the first years of the twenty-first century without referring to the September 11 terrorist attacks on New York City and Washington. That is, in fact, a good place to begin to discuss the new role of information in war.

History will not portray Osama bin Laden as a mere terrorist. Rather, instructors at West Point and Annapolis will cite him as one of the first military commanders to use a new kind of combat organization in a successful operation. One of the most interesting things about the September 11 attack by Al Qaeda and the operations of U.S. military forces in Afghanistan was how the two sides used essentially the same tactics. Consider the parallels:

- Both were connected to their fighters by an encrypted, secure global communications system using a variety of modes—cellular, satellite, fiberoptic, voice, fax, and Internet.
- Both used small teams of special forces covertly deployed deep in enemy territory to assist in terminal guidance. (Of course, "terminal guidance" had a whole different meaning for the Qaeda fighters.)
- Both directed their military operation from headquarters located halfway around the world—bin Laden most likely from Tora Bora, General Tommy Franks from U.S. Central Command headquarters in Tampa, Florida.
- Both used large fuel-air bombs to level high-profile targets and

command centers—BLU-82 "daisy cutters" carrying 12,600 pounds of explosive slurry in the case of the United States, airliners carrying 10,000 pounds of jet fuel in the case of Al Qaeda.

- The leaders of both sides, when facing direct attack, took refuge in underground shelters, as this was the only kind of base that could survive such an assault.
- In Afghanistan, both adversaries were foreigners relying on local allies for ground operations, and both had to compromise their strategic objectives because of that.

Further parallels exist. Both used similar psychological warfare campaigns. Bin Laden disappeared to an undisclosed location, emerging for brief television appearances to boost the morale of his fighters and seek public support from the world's Muslim population. Likewise, Vice President Cheney had his own undisclosed location, and periodically appeared on television to assure the American public that their government was functioning as normal.

I don't mean to suggest for a moment any kind of moral equivalency. The Taliban was a totalitarian, fundamentalist regime that killed apostates, banned music, destroyed priceless antiquities, and treated women as chattel. The United States is a raucous democracy that assimilates all cultures, protects freedom via the Constitution, tolerates both Bach and Eminem, and employs women as fighter pilots and national security advisers. There is no moral comparison.

My point is simply that technology is driving everyone, terrorist and armies alike, to the same tactics. What is more, most of the technology is commercially available, and thus, it is there for the taking. That is why some of the poorest, most backward countries in the world are able to carry out credible military operations against the richest, most advanced countries, using the same methods.

For want of a better name, call this new kind of combat organization a "fighting network"—a way of arranging soldiers for combat. It fits into the tradition of older formations, like the Greek phalanx, the French regimental fighting square, or the NATO armored column. The network has three defining features.

First, it consists of interconnected but autonomous cells. Each cell can operate for long periods independently. But each cell understands the organization's strategic objectives, so each can coordinate its actions with the others and take advantage of local opportunities as they come along.

Second, each cell is armed with extremely potent weapons that make it incredibly deadly. These can be the "conventional" weapons of mass destruction—that is, nuclear, chemical, or biological. Or they can be less conventional weapons, like Al Qaeda has used—hijacked airliners, truck bombs, or exploding dinghies. In the future they might be cheap robotic weapons, like GPS-guided drones.

And third, the cells are linked together by a secure, networked communications system for both logistics and command and control. This system can be a military system, a commercial system, or a combination of the two. Or it can use voice, e-mail, fax, or floppy disks sent by courier or commercial package express. The most important thing is that it allows anyone in the network to communicate with anyone else as required, and does not depend on a single central node.

Fighting networks can be as small as a three-man terrorist organization or as large as a joint task force. They can operate on the scale of a few city blocks or an entire hemisphere. They can use cheap, simple handheld weapons or weapons that cost hundreds of millions of dollars. Their essential feature lies in how they use information technology and how they operate.

Network warfare looks a lot like guerrilla war with incredibly powerful weapons. Or like special forces operating on a global scale. But no matter what you want to call it, what bin Laden accomplished was truly a new, significant step. By using this new approach, the poorest, least powerful country in the world was able to team with a terrorist organization to carry out the most successful military strike ever against the richest, most powerful country in the world—from the other side of the globe.

Consider the effects of the September 11 strike. Al Qaeda killed 3,025 people, representing the greatest loss of life resulting from an attack on the United States in a single day. By comparison, Japan killed 2,403 Americans in its raid on Pearl Harbor, and 2,100 Union soldiers were killed by the attacking Confederate force at the Battle of Antietam during the Civil War (the Confederates themselves lost 1,550 lives).[9] As for the material damage, projected insurance payouts in

New York City alone ranged up to $70 billion. The economic cost to the city was estimated at $83 billion in lost output and the disappearance of 52,000 jobs. Repairs to the Pentagon cost $800 million.

At a broader level, the American economy, already stalling, was knocked further into recession. Washington's political agenda was turned upside down as issues like balanced budgets, Social Security lockboxes, and tax cuts were replaced by homeland defense and bailouts for the airlines (provided $15 billion in assistance). A national medical survey reported that 44 percent of Americans demonstrated a "substantial" stress-related reaction during the five-day period following the attack.[10]

Call it immoral, insane, or plain evil, but the September 11 strike was military genius. What is more, this was not a single isolated attack. It was just one operation in a well-organized campaign that stretched over several years. The first World Trade Center bombing in 1993 killed seven people and nearly collapsed the towers. The August 1998 embassy bombings in Kenya and Tanzania killed 224 people. In October 2000 Afghanistan came close to becoming the first naval power to sink a U.S. warship since World War II—the U.S.S. *Cole,* an 8,200-ton destroyer Al Qaeda bombed in the Yemeni port of Aden.

As we have seen, the United States and Al Qaeda took the same approach to war. That is because many groups can now compete at the same level as many nation-states, and everyone is adopting similar methods, because that is what works. Ever since the Gulf War, anyone who wanted to challenge the United States knew that they needed a new military option. Desert Storm proved that no one can match U.S. forces in a traditional kind of war. The September 11 attacks show that they have found that new option.

The basic concept of how to fight a war has not changed much since Cyrus of Persia created the first organized army about 2,500 years ago. Sometime around 600 B.C. Cyrus organized his warriors into divisions of 10,000 men. Each division was subdivided into battalions, which were in turn divided into companies. It was the first time in recorded history that a central government controlled an army through an orderly, hierarchical chain of command. Lower-ranking officers commanded smaller units, and followed instructions from higher-ranking officers commanding larger units. It was an organization based on the communications of the day: spoken words (or, in the heat of battle, shouted commands).

About two centuries later the Romans added the idea of using professional soldiers instead of citizens and slaves. Roman legions trained full-time and operated from fortified bases. Governments controlling territory had an advantage over nomadic groups or religious movements in preparing for war because they could organize and run an army more efficiently. They could also build defenses around the territory they controlled.

Most armies have been organized more or less the same way ever since, and their basic approach to warfare has not changed much, either. Armies attack an opponent by assembling in formation, and then moving forward on a linear front, mowing down and killing whatever happens to be in front of them. That was the reason for the training and hierarchical command—to enable an army of many men to move as a single unit so soldiers could concentrate their firepower and protect each other.

This approach changed hardly at all for many centuries. In fact, a Roman legionnaire who fought Hannibal's Carthaginian army at Zama in 202 B.C. would probably have felt pretty much at home with a French knight who fought the British at Agincourt in 1415. After the Renaissance introduced gunpowder and the Industrial Revolution introduced mechanization, armies grew larger, faster, and more deadly. But these changes were evolutionary, not revolutionary. Armies still operated as hierarchical, centrally controlled organizations.

The Information Revolution is changing this. Satellite links, fiber-optics, and digital, networked communications make it possible to deliver information almost anywhere immediately. Cheap computers can process even complex data. Everyone in an army can share the same picture of the battlefield (at least in principle), and they do not depend on a central node for all their information.

Now a unit operating alone can size up the situation on the battlefield, take advantage of local opportunities, and coordinate with other units directly. Each unit depends less on its superiors to direct it through every step of a war plan. All of this makes it possible to organize an army as a flexible network of forces, rather than a fixed, rigid hierarchy.

Once you reduce the need for hierarchical command and control, there is also less need to physically arrange armies as hierarchies. This offers opportunities for new tactics. Before, a unit that found itself surrounded behind enemy lines was doomed. Today such a unit may be in the best position to attack its target at its weakest point. Now armies have an incentive to pre-deploy invisibly—even in the midst of enemy territory—exactly as Al Qaeda and U.S. special operations forces did.

The defensive game has also changed. It used to be that the best way to defend yourself was by taking shelter in a fortress. When marching into battle, troops would protect each other by gathering into formation, like the Romans did. Now weapons are so accurate and potent that it is hard to protect any conventional fixed facility, and massing forces merely offers your opponent a better target.

Today dispersion, covertness, and stealth—essentially, information armor—are the only effective protection. What you cannot hide or disperse, you need to bury—which is why so many governments (Russia, Iraq, North Korea, Cuba, Taiwan, Libya, and the United States, just to name a few) have all dug deep, carving out shelters for the leadership and for secret military bases. One report estimates military organizations have built 10,000 underground shelters worldwide.[11]

This was how bin Laden went into combat on September 11. By using computers, satellite communications, and the Internet, Al Qaeda pre-deployed its strike force in America and Canada. Bin Laden controlled the force from halfway around the world; the day before the attack, bin Laden's lieutenants discussed the "big attack" they were about to launch (U.S. intelligence intercepted the messages, but was unable to process them until it was too late). On the day of the attack, Mohamed Atta, the leader of the hijackers, called Ramzi Binalshibh, the planner of the strike, who had fled back to Pakistan. Atta told him, "Two sticks, a dash and a cake with a stick down," a code for "9–11," indicating he was executing the plan, and when.[12]

Each cell in this network could coordinate its action with others, even up to the last minute. While waiting on the tarmac at Boston's Logan International Airport, Atta, on American Airlines Flight 11, used his cell phone to make a call to Marwan al-Shehhi, the leader of the team on United Flight 175. Atta was probably sending al-Shehhi a final "go" signal. At about the same time, back in Afghanistan, the Taliban scattered their small number of jet fighters and tanks into the countryside.[13]

In other words, Al Qaeda and the Taliban were coordinating military operations on a global scale. We might have expected this mastery of networked military action to emerge in our era. Corporations are using telecommuting. Researchers use "virtual teams." Why not the same approaches for an army? Indeed, as we shall see, the Defense Department has been moving in the direction of network warfare for years—we just did not expect to be on the receiving end so soon.

Information technology was the most important feature of the war between the United States and Al Qaeda. Communications networks held both armies together. Communications networks defined the battlefield. Al Qaeda won on September 11 because it had, to use military jargon, "information dominance." It knew where its targets were and maneuvered to attack them. We did not know where Al Qaeda was until it was too late.

The lesson: Today *the ability to collect, communicate, process, and protect information is the most important factor defining military power.* In the past armor, firepower, and mobility defined military power, but now it often matters less how fast you can move or how much destructive force you can apply. Stealth trumps armor, precision trumps explosive force, and being able to react faster than your opponent trumps speed.

If this is true, then *to defeat your opponent, you must first win the information war.* You can do this by making your own information systems more capable, reliable, and secure, or by attacking your opponent's systems so that they are less capable, less reliable, and less secure.

Today the threat that worries people most is terrorism. But the line between different forms of armed conflict is blurring. National armies are planning to use many of the same methods as terrorist organizations. Both organize their forces in similar fashions. Both rely heavily on information technology and on secrecy.

Indeed, as their technology and tactics converge, morality and a willingness to comply with the rules of war might become the *only* difference between terrorist groups and legitimate armies. Governments have an interest in preserving the current international order and thus play by the rules. Terrorists, by definition, want to overturn the existing order. They therefore not only break the rules; they will try to drag down their opponents by forcing them to do the same.

Nonetheless, anyone who uses lethal force today is being driven to similar tactics, because that is what works. The technology is so widely available that all countries will likely use network warfare in some form. The tactics are scalable; the cell in a fighting network can be paramilitary terrorists, special operations forces, or high-tech, mobile forces. They can even be robotic.

Armies will develop these capabilities, find allies who have them, or subcontract for them. Any organization with a global presence and experience in violence could be the next Al Qaeda. It could be an ambitious criminal ring such as the Italian, Russian, Japanese, and Chinese mafias. Or it could be a paramilitary organization affiliated with political parties or movements, such as Hezbollah, Hamas, or the "Real IRA." Narcotics traffickers such as Fuerzas Armadas Revolucionarios de Colombia (FARC) have many of the skills and resources, as do some cults (e.g., Japan's Aum Shinrikyo) and some militant environmentalist and antiglobalization extremists (e.g., the Earth Liberation Front).

Osama bin Laden was a pioneer, and the September 11 strike was a demonstration. The basic ingredients for creating such a lethal network or combat organization are widely available. Many organizations have the global presence and skills required. They will learn the lesson and adopt similar tactics. The threat is not just Al Qaeda fundamentalism, or terror. The threat is a technology combined with an idea. Lethal networks are here to stay.

Chapter 3

"THEY ARE ALREADY AMONG US"

To understand modern warfare, you need to begin, of all places, in Budapest.

There is a joke that you are bound to hear if you spend any time hanging around physicists of a certain age. According to the story, Enrico Fermi posed a question to his fellow Manhattan Project scientists. If, asked Fermi, extraterrestrials with advanced intelligence really exist, why haven't we observed them on earth? Leo Szilard supposedly offered the answer, and the punch line: "They are already among us . . . they are called Hungarians."

Much of the twentieth century's brainpower was born within Hungary: John von Neumann, inventor of both quantum mechanics and the computer program; atomic scientists Edward Teller, Eugene Wigner, Leo Szilard, and Isidor Rabi; Elie Wiesel, humanitarian and recipient of the Nobel Peace Prize; Andy Grove, one of the founders of Intel Corporation. All were Hungarians. Or, more precisely, *ex*-Hungarians. Budapest may resemble a fairyland. Its onion-domed churches and hot-spring fountains may seem enchanting. But the city has always had a dark side. Hungary is at the crossroads of Europe, another way of saying that it has been on the invasion routes connecting Germany, Russia, and Turkey.

Things got especially chaotic in Hungary after World War I, when the Allies carved up the Austro-Hungarian Empire, and got even worse after the Fascists and Communists began to compete for control. Eventually the Fascists won out, and Hungary sided with Nazi Germany during World War II. As a result, the U.S. Fifteenth Air Force flattened Budapest, and the Red Army pillaged what remained. The combination of a cosmopolitan society, a devotion to education, and a penchant for war and riot has made Hungary a leading exporter of human genius.

Among those fleeing all the chaos were Edward and Irene Rona, who sent their son Thomas and his older brother George to the more tranquil environs of Paris soon after Tom was born in 1923. Tom grew up a Parisian, taking his finals in engineering at the École Polytechnique on V-E Day. After a brief detour with the French engineering corps building bridges in Cameroon (partly to dodge an irate father and a shotgun marriage), Rona returned to Paris and met Monique Noel, a bank clerk. Tom and Monique soon married.

Tom and Monique began moving farther and farther west to escape the chaos and constraints of the Old World—first to Montreal, then to a junior faculty position for Tom at the Massachusetts Institute of Technology. By that time he and Monique had three sons and a daughter. An assistant professor at MIT earned an annual salary of $2,400, so money was tight. Tom heard about an opening for a staff scientist at Boeing. He and Monique flipped a coin. Boeing won, and they piled the four kids in the car and began the drive to Seattle. Tom reported to work, and Monique got a job at the University of Washington.

It's hard to tell if Rona ever actually used his engineering degree at Boeing, at least in the sense that he never designed a bomber or missile. Officially he was a senior scientist. Defense contractors like Boeing charge the government an extra fee beyond the basic materials and labor it takes to build a B-52 or an air-launched cruise missile. This money goes into an account for preparing bids and proposals (B&P, in defense contractor jargon), which Boeing used to pay for most of Rona's salary.

The idea was that Rona would develop new ideas for using military hardware—hopefully, new Boeing hardware. In any case, Rona effectively had a license to look at any technology or topic that seemed interesting and was a potential market for Boeing. It was almost as good as being a professor at a major university, if not better.

People were still trying to assimilate the lessons of World War II when Rona arrived at Boeing. The war was just ten years past, and everyone was still trying to figure out the legacy of what Winston Churchill called the "wizard war"—the contest of electronic weapons and countermeasures.

The British had learned the hard way just how complex this game could be. Electronic warfare had become a critical factor in World War II during the Battle of Britain. In 1940 émigrés who had escaped the Continent told British intelligence that the Germans had developed some kind of "beam" weapon. At first the Brits thought this might be a ray gun that could shoot down aircraft by electronically frying their ignition systems; the refugees had talked about the Germans' testing a "beam weapon that stopped cars."

When they dug deeper into the reports, though, the Brits discovered that translators had mangled the syntax; what the émigrés really meant was that the Germans were testing a weapon that required them to stop cars—that is, stop nearby traffic to avoid radio interference. Later the British bugged the cells of some German POWs and learned that the weapon was a radio device for guiding bombers to their targets.[1]

The Luftwaffe had developed a system of steerable radio beams criss-crossing Britain from stations in France and Germany. A primary beam traced a path to the target for the pilot to follow. If the pilot went off course, the beam would grow weaker, and this would trigger a buzz in his headset telling him to correct his heading. Additional beams bisected the flight path to alert the bombardier as he approached the release point. When the signal from the final crossbeam peaked, the bombardier heard a signal in his earphone and he knew he was over the target. It was like an electronic "X marks the spot."

Eventually the British learned how to deduce the direction of the beams and thus figure out which targets to protect. Yet this breakthrough had little to do with detecting the beams themselves. The real trick was in intercepting and deciphering the communications that the Luftwaffe used to tell their radio stations where to aim their beam each night. This revealed the direction and frequency of the beams, and with this information, the British could both determine the target the Germans planned to bomb, and also track a beam back to its transmitter.

The British soon learned how to transmit their own signals on the same frequency, jamming the receivers on the aircraft or causing the bombardier's signal to go off before his aircraft actually reached its target. Combined, all of these separate ingredients provided the components of an information warfare operation. The British learned where the Germans planned to attack, manipulated the Germans' view of the situation, decided how to respond, and then reacted before the Germans knew what they were up to.

The British thus had the advantage, but they then had to answer the Perennial Question of Information Warfare. Even Indians faced the Perennial Question when they first discovered a rival tribe using smoke signals. It still confounds information warriors even today: *Deny, deceive, destroy, or exploit?* Do you transmit your own smoke signals to interfere with his? Do you send bogus signals to confuse your adversary so that he is easier to kill? Do you find the enemy sending the message and kill him? Or do you quietly watch the signals so you know where your adversary plans to be, head him off, and kill him then?

If the Brits jammed the beams or destroyed the radio stations, they might have eliminated the beam system temporarily. The Germans, however, would then have known that their guidance system was successful, or at least successful enough that the British believed they needed to neutralize it. Some German program manager would likely have used this fact to justify a request to build new transmitters that were better hidden and transmitted on different frequencies.

On the other hand, if the British simply sent their fighters to protect the target designated in the deciphered message, some German bombers would have gotten through. Even worse, if a squadron of Spitfires or Hurricanes regularly appeared at the appointed place night after night, the Germans would have eventually figured out that the British had cracked the cipher, which was about the most valuable secret the British had at the time.

This question of how best to attack an information system once you have the advantage—and who should do it—is an old one. Even today it is good for countless interagency meetings and memoranda. The Indian tribes probably had strategy meetings to deliberate how to deal with the smoke signal threat. One warrior likely proposed killing the signal senders, leaving their rivals blind, while another argued passionately for using the information to lay an ambush for the enemy braves who were being directed by the signals.

In the end, the British did a little of each. They jammed some of the signals some of the time, just enough to confuse the Germans. They bombed some of the radio stations. They even leaked some fanciful reports claiming that they had learned how to "bend" radio beams.

The goal was to keep the Germans off balance, and the plan worked. The Germans never lost confidence in their technology and kept using the same beam navigation system. By September 1940 the British had shot down 1,400 German aircraft. German bombing became less accurate. And Ultra, the secret intelligence based on the decrypted intercepts, remained secret until 1974, when the British government itself revealed its coup.[2]

Even so, no one had really given much thought to how all these pieces—jamming, deception, intelligence—fit together. The really important thing was always the weapon—a gun or aircraft. Communications, tracking, and guidance were merely "support systems."

This was all about to change, thanks to other events underway in Seattle. As far back as 1851, when the city fathers convinced Henry Yesler to build his newfangled steam-powered sawmill along the banks of the Puget Sound, Seattle had risen, and fallen, on each new wave of technology. After Bill Boeing built his first seaplane for the Navy in 1917, the new technologies driving the Seattle economy were mainly based on aerospace—first airmail, then bombers, and more recently, missiles and satellites.

In the 1960s a new technology began to drive the Seattle economy: computers. Up to then, computers were scarce and expensive, and could only run one program at a time. Then in 1957 a young mathematics graduate student named John McCarthy proposed a revolutionary idea: time-sharing, or having a single computer run several programs simultaneously. This completely changed the computer business and, by extension, Seattle.

McCarthy, who was visiting MIT on a fellowship at the time, observed that the slowest part of a computer system is always the person operating it. We work at human speed; the computer works at electronic speed. The computer requires just milliseconds to run a typical calculation. McCarthy, later a distinguished professor in computer science at Stanford, realized that if you can collect computer jobs from many users, the computer can electronically rack-and-stack the jobs as they arrive, perform the operations as capacity becomes available, and thus run more or less continuously.[3]

A Teletype—a typewriter that transmits a different electronic signal for each character in the alphabet—made it possible to do all this from miles away. The basic idea for the Teletype had been kicking around since 1909, but it was not until 1931 that the Bell System had introduced it into commercial telegraph service. It was a major improvement at the time, because it eliminated the need for telegraph operators to learn Morse code; now they could just type a message. By 1968 Bell Lab engineers had adapted Teletypes to send and receive data from time-sharing computers.

It did not take long for some early entrepreneurs to put all the pieces together and see a business opportunity. One could buy a computer, hook users into it with Teletypes, and sell portions of the computer's capacity to companies that couldn't afford to buy their own. Time-sharing became popular at universities, which typically had two or three large computers located somewhere on campus, and tens of thousands of faculty and students who wanted to use them. So it was little wonder that many of the people who tried to get into the time-sharing business were college staff, like Monique Rona.

By then Monique had expanded her repertoire from oceanography to computer science. Soon she found herself running the university's computer center. Tom helped with the financing and Monique got together with some university colleagues to buy a Digital PDP-10. They set up the Computer Center Corporation—C-Cubed, for short—in an office near the campus. Monique put a Teletype on the kitchen table, and they were in business.

The PDP-10 had just hit the market. Being a new machine, it was prone to electronic burps and hangfires. The C-Cubed partners thought they might cut a deal where Digital would give them a discount on the computer if they worked out the bugs.

As it happened, one of the Rona boys attended the Lakeside School, a local private academy, and had heard of some classmates who liked to work with computers. Some of the Lakeside moms had bought their kids a Teletype and a few thousand dollars of computer time from local companies with money raised from a rummage sale. The moms hoped their kids might learn a few computer skills writing programs to play ticktacktoe and the like.

The kids had other ideas. They began playing with the Teletype day and night, taking apart programs and writing some of their own. In no time they burned through the computer time their mothers had bought them and had to look for some new benefactors, just when Monique was looking for some eager minds to test her PDP-10. The boys cut a deal with C-Cubed: the boys would look for bugs in the PDP-10, and C-Cubed would give them some time on its computer.

Two of the kids, Bill Gates and Paul Allen, seemed to have a knack for computers. When the machine crashed, they would fetch the "core dump" from the trash and search through the machine language line by line to find the bug. They wangled operating manuals from the staff. Eventually the "Lakeside Programmers Group" came to know the insides of the PDP-10 about as well as C-Cubed did. So, when they ran out of their allotted time on the machine, they naturally took the simple expedient of fiddling with the computer's operating system to set back the clocks. Bill, Paul, and their Lakeside buddies Ric Weiland and Kent Evans were four of the earliest computer hackers.[4]

Alas, the market for computer time-sharing never worked out as well as Monique had hoped, and C-Cubed eventually went broke, another case of roadkill on the path to IT riches. Gates and Allen went on to other ventures. One was a new company, "Micro-Soft," which they incorporated in April 1975 to sell programs for the Altair 8800, the first personal computer.

Meanwhile, Tom and Monique had that Teletype on their kitchen table. Computers—especially computers that talked with other computers—were part of the family. Tom could see that the Teletype in the Rona kitchen communicating with the PDP in the C-Cubed office near the university was no different than, say, a radar off the coast of New Jersey communicating with a central computer at the North American Air Defense Command in Colorado.

Rona could see that any widget that collected, moved, or processed data was part of a system, each dependent on the other. He could also see that every widget in a system was a potential point of vulnerability, as was the information that passed through them. These so-called support systems were potentially a better target than the weapon itself.

This had not occurred to anyone before, probably because information machines had never routinely talked to one another before. But what Rona now saw was clear: Control the information flowing through your adversary's computers and communications networks, and you could control the outcome of a battle or a war. Or, as Rona's monograph, *Weapons Systems and Information War,* put it, "Countermeasures aimed at the external flow of information will be further improved to the point that they may well become crucial in influencing the outcome of future engagements."[5]

Often, in common English, this meant that the best way to defeat your enemy was to attack the components of its information systems, and Rona was broad-minded in defining "component." It included the hardware, of course, but it could just as easily be software, the people operating the system, the people getting the information out of the system, or the data that traveled through it. The best component to attack depended on the opportunities at hand and the risks one was willing to take.

In fact, Rona said, information war offers a major advantage over the old-fashioned kind. In information war, you have a menu of options—that Perennial Question of *deny, deceive, destroy, or exploit?* It offers an entire matrix of potential pressure points and methods for attacking the enemy. And if your target happens to figure out what you are up to, you can change tactics. The only challenge is to adapt faster than your adversary.

Boeing published Rona's monograph in the summer of 1976 as a "think piece" for company staff and customers. Tom Rona was the first person to use the term "information war" in print. Considering that the Internet was still thirteen years away and a "home computer" was something you built from a box of components with a screwdriver and a soldering gun, it was not a bad piece of prognostication.

Chapter 4

THE ASYMMETRIC WARRIOR

Many people think that information warfare—in particular, the idea of attacking computer networks—was the Defense Department's version of the dot-com boom. Some even believe that information warfare is like the management theories of Enron or Worldcom, a fashionable notion that will not last. But that is a faulty reading of history.

The origins of computer wars have a lot more to do with the Cold War of the 1980s than with the browser wars of the 1990s. The drafters and planners for information warfare were mostly Navy and Air Force officers who worried about fighting a nuclear war with the Soviet Union. Even the rationale for information warfare was originally a grand strategy aimed at toppling the Soviet empire.

To understand all this, one needs to go down the Pacific coast from Tom Rona's Seattle to Santa Monica, California, where the Air Force set up Project RAND in 1946 as one of the first not-for-profit think tanks. Project RAND was partly a leftover from World War II and partly preparation for the Cold War.

Planning a strategic bombing campaign is a lot like planning public works in reverse. Instead of building infrastructure to promote economic growth, the goal is to destroy infrastructure to kill an economy. The methodologies are even similar. You can run an industrial flow analysis to identify a manufacturing choke point and make a factory more efficient. Or you can run a flow analysis to identify a single point failure that will cause the whole plant to shut down when you blast it to bits.

From an analyst's point of view, this is really interesting work. Dismantling a country's infrastructure from 30,000 feet is usually a problem with several potential solutions: How many bombers? What kinds of bombs? Which targets? What is the optimal attack sequence? During World War II the Army Air Corps recruited—that is, drafted— economists, engineers, mathematicians, and assorted big-picture thinkers to analyze these problems.

After the war General H. H. "Hap" Arnold, the soon-to-become chief of the soon-to-be-created U.S. Air Force, knew he would need the same kind of eggheads. Unfortunately, the Air Force would have a hard time competing with universities and corporations for this talent in peacetime. And, truth be told, they were an odd fit for a uniformed service in any case.

That was the reason for Project RAND (Research and Development). At first, the Air Force put Project RAND inside the Douglas Aircraft Company. The operation got started just a month after the end of the war, as RAND set up shop in the Douglas Aircraft plant in the Santa Monica municipal airport.

The Air Force soon realized RAND would analyze proposals for all kinds of Air Force projects—including, most likely, aircraft and missiles proposed by Douglas Aircraft. Everyone knew this was an obvious potential conflict of interest. So in 1948 the Air Force spun off Project RAND into a nonprofit corporation. The new company continued to grow, and it soon moved to an office complex two blocks from the beach—a stone's throw from the Santa Monica Pier and conveniently situated next to Chez Jay, a local bar and hideaway.

Andrew W. Marshall was one of the early hires. Marshall started out publishing papers with ferociously academic-sounding titles on basic statistical theory ("Some Tests for Comparing Points of Two Arbitrary Continuous Populations") and methods for assessing the effectiveness of bombing operations ("The Estimation of Parameters in a Physical Vulnerability Model"). Soon, though, Marshall began to focus on bigger questions, like how to measure the military capabilities of the United States and the Soviet Union and, by implication, how to tilt the balance in our favor.

Marshall soon found a theme that he would return to repeatedly over the next fifty years: "asymmetric threats." At the time, most people measured the so-called superpower arms race by comparing apples to apples. Marshall saw that such comparisons were usually irrelevant because you usually needed to compare apples to oranges. Comparing the number of tanks the Soviets had to the number of tanks we had told you nothing, said Marshall. Neither did comparing the size of their missiles to the size of our missiles. The reason was simple. Countries playing offense need different weapons from countries playing defense.

Think of it this way: If you are attacking a castle, you need ladders to climb walls, battering rams to smash in the front gate, and shields to protect you from the arrows that the defenders will shoot at you from the parapets. If you are defending the castle, you need arrows to shoot down at the attackers, vats of boiling oil to pour on their heads, and so on.

So it doesn't mean anything to count the number of siege ladders on each side. This may seem painfully obvious today, but it was new for many politicians, generals, and pundits who were accustomed to thinking the strategic balance was something you read off a scorecard to see who was ahead.

To complicate matters further, the Soviet Union and the United States had different strengths to draw on. For instance, the Soviets could build bigger missiles because they had better rocket engines. But U.S. missiles were more accurate because we had better guidance systems. So a missile-to-missile comparison was misleading.

The most interesting thing about asymmetric threats, though, was that even the strongest army or air force could have an Achilles' heel. Indeed, a country might create such vulnerability for itself and not even know it.

This was a point that eventually made Albert Wohlstetter famous. Wohlstetter, one of Marshall's colleagues, described the ultimate asymmetric threat, in which a critical flaw, astutely exploited by an opponent, could leave even the most powerful military force dangerously weak.

At the time, the U.S. Strategic Air Command had four or five times as many long-range bombers as its Soviet counterpart. However,

Wohlstetter noted that most U.S. aircraft were parked out in the open on airfields or, even worse, at foreign bases that were within range of a pre-emptive strike by the Soviet air force.

If the Soviets got the first punch, Wohlstetter said, they could destroy most of the U.S. bomber fleet. In fact, our vulnerability, combined with their own weakness, gave them an incentive to strike first if U.S.-Soviet relations went sour. This was really perverse, because the whole point of U.S. nuclear forces was to deter a Soviet attack. Wohlstetter claimed that, as configured, the U.S. force was not only failing to deter; it might even provoke a Soviet strike.

Usually the best army for winning a war bore little resemblance to the army that it had to defeat. They were, in other words, asymmetric. The ultimate goal, said Marshall, was finding the links in an opponent's forces that were not only weak, but which he couldn't fix even if he tried. Now *that* was really an asymmetric threat. It was the way to win not just a battle but the whole superpower competition.

Marshall knew that the idea of asymmetric threats was not only correct, it was impeccably logical. In fact, it was elegant. If you were defending, it revealed the vulnerability that others had overlooked. If you were on the offense, it could lead to a new strategy that was irresistibly efficient.

Marshall continued to work on measures of military capabilities until 1971. By then he had been at RAND for more than two decades. It wasn't a great time. The United States was en route to losing the Vietnam war, and it seemed that everywhere you looked in the developing world, some country was setting up a Marxist government or signing a "Friendship and Cooperation Treaty" with the Soviets. The United States badly needed a strategic solution—elegant, efficient, or otherwise.

The Soviet Union had pulled even in the superpower competition—or ahead, depending on how you measured it. Of course, measuring and comparing military capabilities happened to be Marshall's specialty. So, he asked, even if the situation looks bad now, can the United States plan a strategy that wins in the long run? To borrow a term from the economist David Ricardo, the United States needed to find its "comparative advantage," the areas in which it had a natural edge over the Soviets.

Computers seemed promising. A totalitarian country like the Soviet Union had to control the flow of information. The Soviet regime was so worried about the threat of *samizdat* that in 1971 the average citizen could not get near a Xerox machine, never mind a computer. This was a weakness that would help destroy the Soviet Union, although most Soviet leaders did not appreciate it at the time. The Kremlin's information technology phobia allowed the United States to plot a strategy that would emerge in full bloom during the Reagan administration and eventually push the Soviet Union over the edge.

Because the Soviets were so obsessed with controlling information, they could never have their version of Bill Gates and Paul Allen. The very idea of two teenagers foraging for free computer time on some company's PDP-10 was utterly nonsensical in the Soviet context. There were no computers in the private sector. For that matter, there was no private sector. All of the computers in the Soviet Union resided in state-sponsored research institutions and military bases, where they were protected behind security perimeters defined by fence lines three-deep.

The business world was beginning to outpace the government in developing new information technology. Marshall could see that as long as the Communists remained communists, the Soviets would never catch up. The United States, said Marshall, had to figure out some way to use this advantage.[1]

As it happened, James Schlesinger—a RAND alum himself—became Secretary of Defense two years later and found himself searching for someone to head an "Office of Assessment and Strategic Planning." Schlesinger had discovered he had no one in the Department of Defense with the responsibility to step back now and then, look at the world, and estimate who was ahead, who was behind, and suggest what the Defense Department could do about it. He called his old colleague Andy Marshall, and Marshall moved to Washington.

"When we brought the office here, we concentrated on the assessment part and left off the strategic planning," Marshall recalls, smiling because he knew which part would likely have the greater influence. Thus began the Office of Net Assessment.

"We got hold of some classified writings about how the Russians calculated the correlation of forces," he continues. "We were interested in reconstructing their calculations so that we could see how we might influence them." In other words, Marshall planned to find out how the Soviets decided whether they were ahead or behind in the superpower competition, and then mess around with their minds.

The Soviets seemed especially interested in computers and communications systems and how to attack them. Soviet writers recalled how the Red Army prevailed in World War II through *radioelektronaya bor'ba,* "radio-electronic combat."

Censorship was so strict in the Soviet Union that Soviet military writers often had to use analogies and examples from the Great Patriotic War to make their point. According to the Soviets, there were four basic techniques in radio-electronic combat: jamming an opponent's communications, using radio direction finding to direct artillery that would destroy the enemy, concealment, and inserting *disinformatzia* into the adversary's circuits to confuse him.[2]

The Soviet writers were worried. According to their calculations, U.S. soldiers were better *radioelektronaya bor'ba* warriors than they could ever hope to be. They had taken a look at the American electronics industry and the systems that the Defense Department had deployed, and concluded that U.S. war planners must have given the topic a lot of careful thought.

As it turned out, they were mistaken. "They thought that we were going to attack their command and control systems, even though our capabilities to do so were really not well developed," Marshall recalls. But this didn't matter. "The important thing was that *they* thought we were in a radio-electronic revolution, and then screwing them."

So Marshall began in 1978 to draft a Net Assessment on Command and Control. "The more we looked at it, the more we could see that the center of the change in warfare was in the sensors and the information systems. The information aspect of weapons had always been important, but now it was more important than ever. So one of your key operational goals had to be to establish an advantage," Marshall recalls.

Before long Marshall heard about Rona's monographs on "information war" at Boeing. The two met and hit it off. Marshall could see Rona was onto something. "No doubt about it. Tom was the first person to put it all together," recalls Marshall. For the next four years

Marshall supported Rona's work on the theory and practice of information war, or, as it came to be called, information warfare. It was a bargain, as Boeing still picked up half of Rona's salary.[3]

As one connection led to another, Rona himself came to Washington a few years later. Ben Plymale, a vice president at Boeing and former defense official from the Nixon administration, was Rona's friend and mentor. Plymale introduced Rona to William van Cleve, a professor at the University of Southern California. Van Cleve had been an adviser to Ronald Reagan during the 1980 campaign and was directing the Defense Department transition team for the new administration. Rona started working for van Cleve, and soon caught the eye of William Graham, who was also working on the transition and was about to become Reagan's new director of the White House Office of Science and Technology Policy. Graham offered Rona a job as deputy director.

After Rona's stint in government, he remained the rest of his life as a Washington writer, consultant, savant, and bon vivant to an informal band of mid-level officers and officials on their way up. In coining the term "information warfare," Rona defined the most important factor in modern military thinking. And it was all because he could not afford to be an assistant professor at MIT, and Boeing was willing to pay him to think of new big ideas that would eventually sell more airplanes and missiles.

If anyone had told Tom Rona that his inventing information warfare was originally just a hook for selling Boeing's electronic widgets, he probably would have appreciated the irony. After all, the power to shape the actions of a military organization is the whole idea of information warfare. And that was exactly what Rona did. But if you caught Tom Rona in a quiet moment, he would likely say that he would be satisfied if he had just discovered some way his adopted country could avoid all the chaos that he had seen in his own life.

Chapter 5

SHOOTING DOWN A MIG

At about the same time Tom Rona was showing why a modern war-rior should focus on attacking his enemy's information systems, an Air Force pilot was showing how. The connection was almost as unpredictable as Tom Rona's.

Harry Hillaker recalls a story that made the rounds during the Vietnam war. Hillaker, now retired in Fort Worth after a career at General Dynamics as an aeronautical engineer, was the chief designer of the F-16 Fighting Falcon.

According to the story, in April 1965 four Air Force F-105 Thunderchiefs—"Thuds"—were on a mission to take out the Thanh Hoa—"Dragon's Jaw"—Bridge in North Vietnam. Seemingly from nowhere, a flight of MiG-17s dived into the formation. They immediately shot down two of the American jets. A third F-105, damaged, tried to escape. The pilot of the fourth F-105, trying to cover his buddy, discovered that yet another MiG-17 had managed to get behind him.

The F-105 was one of the Air Force's newest aircraft, part of the Century Series, the second-generation jet fighters designed in the early 1950s. The MiG-17 was designed to replace the MiG-15, which had flown in the Korean War. It had better handling than its predecessor. But it was much, much slower than the Thud. With a top speed of 1,300 miles per hour, the American jet could go twice as fast as the MiG.

Unfortunately, that speed was little help now. To hit top end, the

Thud would need to light its afterburners, and they would suck down so much fuel that the aircraft would never get home. The Thud's speed was only good for short bursts, mainly in bombing runs. The F-105 would have to win a dogfight with the MiG to escape.[1]

The American pilot tried one move after another. But each time he turned, the MiG turned faster and sharper. The MiG, in effect, was able to cut the corners. With each successive turn, the MiG pilot's advantage added up. After several turns, the MiG had maneuvered so that it was almost directly on the tail of the American—in the "kill position," where an aircraft can shoot at its target, and its target cannot shoot back.

Running out of options, the F-105 pilot recalled a briefing the day before from a captain visiting from the Fighter Weapons School, the Air Force's combat training program at Nellis Air Force Base in Nevada. The captain had been describing defensive tactics, and one move sure worked for an F-105 in extremis: pull the stick hard back and left, push in the left rudder pedal, and snap-roll the aircraft.

At first glance, this move did not make sense. Everyone knew that the F-105 was a big aircraft and rolled like a pig. In fact, that was exactly why the American was in his current predicament; the MiG could roll and turn so much faster. But, with nothing to lose at this point, the pilot gave it a try.

The F-105 was such a pig, in fact, that when the pilot snap-rolled it, the aircraft basically turned into an aerodynamic brick. It decelerated with such a vengeance that the MiG, at that instant faced with the odd problem of being *too good* an aircraft, went flying by.

The American was so startled by the effectiveness of the maneuver that he wasn't prepared to pull his trigger even as the MiG passed through his gun sight. Of course, missing a kill was not high on his list of concerns at the moment. Free of his pursuer for the necessary instant, the pilot made his escape.[2]

The captain from Nellis was John Boyd. Read any account of Boyd's life, and you quickly come across the story of "Forty Second Boyd." Boyd offered a standing bet. He would let any pilot start an engagement on his tail, and within forty seconds, Boyd would reverse the situation so he was in the kill position, sort of a fighter pilot judo flip. If he failed, he would pay the challenger $40. Boyd never lost the bet.[3]

Boyd, who was assigned to the Fighter Weapons School to teach combat tactics, had figured out the basic moves that gave any particular aircraft its best path to outmaneuvering an opponent. He collected his ideas on air-to-air combat in the *Aerial Attack Study,* which became a basic primer of tactics used both in the United States and abroad.

Boyd might have remained just a barroom legend in Air Force officers clubs, fading as one generation of pilots replaced another. He did not, however, because he was born with the nonconformist gene. Boyd was unhappy unless he was upsetting some kind of conventional wisdom. The particular orthodoxy that Boyd was taking on at the time was the importance of speed.

Speed had defined the American fighter plane in the 1950s. Indeed, at the time, the purpose of developing new aircraft was mainly to produce faster aircraft. Faster fighters were presumably better fighters. Boyd thought this was nonsense. Experience told him otherwise. To make his point, he collected data and showed that "obsolete" Soviet-built aircraft like MiG-15s and MiG-17s—slow, stubby jets designed in the 1940s—were beating their American competition in Vietnam. The reason, argued Boyd, was *transients.*

As the F-105 pilot had discovered, the MiG could change direction much faster than its competition. So when any American jet and a MiG tried to match each other turn for turn, eventually the MiG would wind up on the tail of the U.S. fighter. Boyd's point was that, in most situations, *absolute speed is much less important than the ability to move from one state to another.* In other words, transients.

Boyd called his idea the "energy maneuverability model." It summarized the performance of a jet as the sum of its transients—in this case, how fast an aircraft could change its velocity or rotation rate at any point in its performance envelope. A jet's performance score depended less on its top speed than on its ability to turn and accelerate.

Boyd had a problem, though. He was assigned to Eglin Air Force Base. Carved out of the Florida panhandle, Eglin was really just a huge expanse of palmetto and scrub pines the size of Rhode Island that the Air Force used as a test range. Boyd was paying his dues for the Air Force's paying for his industrial engineering degree from Georgia Tech. Eglin was not in the business of designing aircraft, and so Boyd had no business analyzing aircraft designs.

Undeterred, Boyd "borrowed" the required computing time with the assistance of Tom Christie, a civilian mathematician. The two struck up a friendship. Boyd then borrowed some data on Soviet jets in a similar fashion from some ex-Nellis friends who had been assigned to the Air Force's Foreign Technology Center at Dayton, Ohio. FTC was the Air Force's intelligence office responsible for "system threat assessments"—that is, analyzing the opposition.

Boyd and Christie put in the performance characteristics of leading American and Soviet fighters and calculated their overall maneuverability. The results showed that as often as not, the Soviet jet had an advantage over its American counterpart. At any given moment, there was some maneuver that would give the Soviet jet an advantage of a step or two.

In the real world, this meant that as the Soviet and American aircraft made move and countermove in a dogfight, the sum of the maneuvers would eventually favor the Soviet aircraft because it had better transients. This was even true of the new fighter that the Air Force was just beginning to develop. Boyd applied the energy maneuverability model to the proposed designs. The results showed that most would not be able to defeat their competition in head-to-head combat.

This exercise got Boyd a lot of attention from his superiors. Anyone in the military understands that "getting a lot of attention from your superiors" is not necessarily a good thing for an Air Force major. The military needs unconventional thinkers, but unconventional thinking runs counter to the military's chain of command and tradition. It's much easier to follow the status quo, and most officers do. The exceptions are kind of odd, brilliant, cagey, or manic—or, as in the case of John Boyd, all four.

Boyd framed the issue as a question that was impossible for Air Force officials to duck: "How can the Air Force design a fighter that will not be able to beat its competition in a head-to-head dogfight?" Besides the fact that Boyd had data on his side ("Why are our fighters getting shot down by a bunch of Third World pilots?"), no official could argue the opposite—that it was okay if the newest, best U.S. fighter would only be second best in a dogfight.

Boyd's argument had the effect of stripping out extra features that would have added weight and complexity. The new aircraft, the F-15 Eagle, was the first Air Force fighter in over a decade designed specifically to outmaneuver the enemy aircraft it would meet in combat. Thirty years later, armed with its superior transients, it remains in service, and is still considered one of the best fighter aircraft in the world.

Boyd extended his model and discovered an interesting fact. In a dogfight, maneuverability often mattered less than simply which pilot saw his opponents first. The pilots of the two F-105s lost over the Thanh Hoa Bridge could attest to that fact; they probably never even knew the MiGs that shot them down were there.

The reason was that U.S. aircraft were easy to see and hard to miss. The Century Series aircraft were all big, and their engines put out a lot of smoke. A MiG pilot spotted his opponent way before the American could. In effect, the MiG would often win a dogfight before the American even knew he was in one. Realizing that there was an opponent nearby in what seemed up to then to be clear blue sky was the first, and often the most important, transient of all.

Boyd eventually reduced his principle down to an acronym for the four steps required for outmaneuvering, and defeating, an opponent: OODA (observation, orientation, decision, action). Gather data about your situation. Evaluate the data against your existing knowledge and your objective. Choose a course of action. Execute. The pilot that could complete the "OODA loop" faster than his opponent—that is, the one with better transients—would win.

But Boyd understood that the OODA loop was not just about dogfighting or, for that matter, the turn rate of an aircraft. Every step in the OODA loop depended on collecting, processing, or moving *information*.

To see why this is so, suppose your aircraft can turn, say, 10 percent faster than your opponent's. Or 20 percent. Whatever your advantage, it means nothing until you actually begin a turn. If your opponent can begin turning before you have completed collecting and processing information, then, for that instant, your advantage is zero. In fact, it is as though you have not decided to move at all. For that instant, your opponent has a maneuverability advantage that is infinite.

This was the real answer to the Perennial Question of Information Warfare. You can collect, analyze, and move your information faster than your opponent to get an edge. Or you can cut off your opponent from his own information sources, distort his processing, or prevent him from issuing commands. You can fight the information war inside a weapon's circuits, or inside a commander's head. There is no single approach that is always best, but the ultimate objective is always the same: collect, process, and apply information faster and better than your opponent. Whoever gets to the end of his OODA loop first gets to take the first shot. In modern warfare, that's often the only shot. *Information warfare is whatever you need to do to get to the end of your decision cycle before your opponent gets to the end of his.*

Boyd—gadfly and iconoclast—retired from the Air Force in 1975. Pilot skills notwithstanding, there was no way he would ever make general. He refused to back down. He went to work as a civilian for the Defense Department, going on the payroll for one day every pay period just to maintain his security clearance. His new project eventually became the F-16, which took the concept of the F-15 even further: minimum transients, maximum field of view.

Boyd also wanted the new aircraft to be affordable enough that the Air Force could buy swarms of them. Boyd believed future air combat would require lots of agile aircraft, flown by pilots armed with an information advantage that would enable them to assess the situation and find the best opportunity to attack. In a way, this was an aerial version of the kind of fighting both Al Qaeda and U.S. forces would use thirty years later—small, autonomous warrior cells, beating their opponent by using information better. They would know where they were, and the enemy would have difficulty finding them. When the opportunity appeared, they would kill.

Boyd became something of a celebrity during the 1980s, but never bothered to cash in. He had always been more interested in asking big questions than in making money. Chuck Spinney, an Air Force captain whom Boyd took under his wing, recalls that after Boyd saw his model of energy maneuverability put to work successfully in the design of the F-15 and F-16, he became preoccupied with a question. Why did he, a fighter pilot with little technical training, come up with the idea? "He became obsessed," recalls Spinney. "He started reading everything he could find that was related to it."

Boyd became a scholar of the art of warfare, and as he got ever more immersed in his later years, it was sometimes hard to know on which side he fell of that fine line that divides charisma from cult leader. The World According to Boyd was a mix of warrior Bushido, a penchant for history, a bit of cost-benefit analysis, all wrapped together in an enveloping disdain for bureaucrats, cronyism, and going along to get along.

Boyd did not have colleagues; he collected disciples, and his Fighter Mafia was just that: a close-knit family of officers, officials, and reporters where personal and professional relationships blurred. Spinney takes exception if you call Boyd his surrogate father—he had a great relationship with his own. But when *Time* magazine came with a once-in-a-lifetime opportunity to be on its cover, Boyd waved it off. He pushed Spinney forward for the cover shot instead, and in the process made the young analyst the public face of military reform, giving his protégé a career insurance guarantee that Boyd did not want or need.[4]

A late-night telephone conversation with Boyd would typically begin at ten o'clock to discuss wing loadings and end at two o'clock with the ancient Chinese military thinking of Sun-tzu. Some of the entourage put in a "Boyd line" at home just to keep a phone free for their families to use. Boyd might be hard to take, but you could not stay away; he was unbearable and irresistible all at once. And there wasn't much room for compromise. For instance, the F-16, a member of the Fighter Mafia would say, might be a great aircraft, but it was never as good as it could have been because bureaucrats "polluted" the design. Dogma intersected with public interest.

When Boyd retired for good in 1988, he moved to Florida and began to think about how his ideas applied not just to aerial combat but also to armies, organizations, businesses, and life in general. In Boyd's mind, life was one OODA loop stacked on top or embedded inside another.[5]

A basketball player who can process his OODA loop faster will be one step ahead of his opponent. Armies that complete their OODA loop faster are the ones that get to the scene fastest with the mostest. Companies that can OODA faster than their counterparts will see opportunity, focus their resources better, and eventually drive their competitors out of business. The idea of a decision loop has percolated through science, business, and society so thoroughly that most people who use the term don't even know that it came from a theory about how to shoot down MiGs.[6]

But, contrary to the saying, ideas *can* exist in a vacuum. To understand how information war became a part of how American forces plan to fight, we need to look at how the Department of Defense works.

Chapter 6

THE PENTAGON LABYRINTH

Navy officers call Newport, Rhode Island, "hardship duty"—with a wink and a nod, because everyone knows it is anything but. If you drive down from Providence, you get to cross Narragansett Bay and view the harbor via the Mount Hope Bridge, a particularly attractive suspension bridge from the late 1920s. Cut out a few extra hours, and you can tour the mansions along the coast, dating back to when Gilded Age robber barrons competed for who could build the most opulent replica of a French château or an Italian villa.

A century ago, the Navy developed torpedoes at Newport, and during the late 1900s it was the home port for dozens of destroyers. But these days Newport is mainly in the ideas business—engineering, education, and thinking about the future of war. The Naval War College, which takes up a good portion of Naval Station Newport, has an entire department devoted to war gaming, and it is a popular venue for meetings when everyone wants to get out of the Washington mind-set.

A couple of years ago a friend invited me to a meeting at Newport. The attendees were looking at the Navy's future role in "space control," or operations to protect U.S. satellites and neutralize hostile satellites. The idea was to get about three dozen people—mainly Navy officers, plus a few top civilian officials—all in the same room to kick around some ideas.

It was fascinating to watch. Some junior officers—lieutenant commanders, commanders, plus one or two captains—were gung ho. They couldn't get the Navy into space control soon enough. They worried about foreign countries that would soon have satellites that could track their ships or detect their communications. But they were also just a bunch of ambitious young men and women who thought space control was neat. They joined the Navy to find a challenge, and this was new turf.

Then one of the admirals took the floor. Sure, he said, it was great to talk about new missions. But he had to make sure the Navy could carry out the missions it was currently assigned. The defense budget does not change much in any given year. So, before anyone proposed new activities for the Navy, he wanted to know which missions the junior officers were prepared to write off. That is, tell me which existing program you want to cancel.

You could just feel all the air being sucked out of the room. The younger faces became long—partly because they knew the admiral would have a hand in approving any project they proposed and, possibly, their own promotions. But it was also the way he put the argument. No one would stand up in a meeting like that and tell another officer to his face that his job is obsolete and should be cut. The discussion was over.

It is easy to think of the U.S. Department of Defense as a huge 2.1-million-person monolith operating from the Pentagon across the river from Washington. But in reality, the Defense Department is a federation of many separate organizations—some very large, some small—spread around the world.

The politics and the separation of powers in the Defense Department often resemble the politics and separation of powers in the U.S. government. The Pentagon has its own checks and balances, and its own culture. The art of maneuvering an idea through this labyrinth of offices and officials requires a special set of skills.

The U.S. Department of Defense can be understood as having three parts. The first part is the Office of the Secretary of Defense (OSD), which itself consists of several organizations. Andy Marshall's Office of Net Assessment is part of OSD. So is the Office of Program

Analysis and Evaluation, (PA&E), the organization that hired John Boyd after he retired from the Air Force.[*]

OSD is the Defense Department's corporate office, similar to, say, Boeing's headquarters in Chicago. Like most corporate offices, it has loads of power and influence, but most of it is indirect. It can draft policies and develop ideas, but it depends on others to put them into practice. It approves budgets, but it depends on others to do most of the initial drafting. It is the main link to the President, but it depends on others to carry out the department's main function—fighting wars.

The uniformed services—the Army, Air Force, and Navy (which also includes the Marines)—are the second part of the Pentagon. Under the law, they are responsible for "generating forces," that is, training people and buying weapons. To continue the metaphor, they are analogous to Boeing's aircraft plants in Los Angeles, St. Louis, and Seattle. Each makes aircraft, spacecraft, or missiles under the parent's name. But each was created separately, each was an independent company at one time, and most important, each has a history, tradition, and culture all its own. They also often compete with each other, with both the benefits and the complications such competition entails.

The uniformed services do not like to serve à la carte. They deliver combat-ready units: armored divisions, naval battle groups, and bomber wings trained to carry out operations according to their ser-

[*] People often make light of the military's habit of turning just about everything into an abbreviation or acronym. Often books about defense have a standard disclaimer in the preface: "We have tried to keep the technical jargon to a minimum, etc. etc."

That is a mistake. Military jargon reflects an important development in national security policy: the growing distance between the military and the public. The defense community, and especially the uniformed services, have become a separate society. Any sociologist will agree that a distinctive language is a classic mark of whether a group constitutes a separate, bona fide society with its own culture.

Thanks to technology, armies, navies, and air forces today require fewer people. Yet they also need people to remain in the force for years. Otherwise, training costs balloon, and units lose the opportunity to learn how to work as a team. The problem is that, as the military becomes more professional and specialized, it also grows more distant from civilian society. Most Americans today know little about how their armed forces work. To make matters worse, this is happening at a time when we need cooperation more than ever between the military and the public.

So I've tried to minimize the jargon. But try to take in the culture. It's akin to learning how to order a particular vintage from a French wine list and impressing the sommelier. Or, as they would say in the Army, *Hooah.*

vice's doctrine for fighting wars. Although the military reforms of the 1980s require officers to serve in "joint" assignment to earn promotion, most officers spend their first ten or fifteen years almost entirely within a single service, and get steeped in that service's culture.

The third part of the Defense Department is the fighting arm of the Pentagon—the part that actually commands U.S. combat forces. The Pentagon divides the world into nine parts—five geographic regions (the Pacific, Central Asia, Latin America, Europe/Africa, and North America) and four functions (special forces, nuclear forces, transportation, and the development of new forces). A four-star general or admiral—a "combatant commander"—is in charge of each.

These commanders are the people who draft U.S. war plans and direct U.S. forces in combat. Until recently combatant commanders were called "commanders in chief," or CINCs (pronounced "sinks"), a term that dated from World War II. Many people in the Defense Department still seem to call them that, since over the course of fifty years the acronym had morphed into a word, and most people probably don't think about what it stands for. Everyone knows the Commander in Chief is the President.

To complete the metaphor, think of the CINCs as being comparable to American Airlines or United. They are, in effect, the customers for the Army, Navy, and Air Force. Boeing delivers airliners; United schedules the routes. Similarly, the Air Force delivers a fighter wing; a CINC decides where to send it. In each case, the customers have much influence, but little direct control over the products.

The chain of command leads from the President through the Defense Secretary to the CINCs, but there is another player in this mix who is very important: the Chairman of the Joint Chiefs of Staff. Like the CINCs, the Chairman is also a four-star general or admiral, but, by law, he is the highest-ranking military officer in the Defense Department. His job is to see that the services work smoothly together, and he is also, by law, the President's military adviser. The Chairman has his own supporting organization, the Joint Staff.[*]

[*] Just to add one last bit of terminology and complexity: Collectively, the heads of the Army, Navy, Air Force, and Marines, plus the Chairman and Vice Chairman, comprise the Joint Chiefs of Staff (or "Joint Chiefs," for short, and not to be confused with the Joint Staff). The Joint Chiefs do not run anything, but exist mainly just to get the heads of the

All of this may seem like an overlapping, ambiguous organization, but it has a sort of creative tension—again, similar to what one finds in the American political process. The President can tell a combatant commander directly what he wants done; but, to ensure that the professional military's opinion gets heard, the Chairman of the Joint Chiefs outranks the CINCs and can go directly to the President. The Army, Navy, and Air Force can develop their own technical expertise and culture (both important for winning wars). But they cannot send their forces out to fight alone; the CINCs make sure that there is an integrated plan.

In any case, no matter how arcane the structure of the U.S. Department of Defense may seem, no one else in the United States is in the business of fighting wars—with the possible exception of the Central Intelligence Agency (or simply, the "Agency"; no one calls it the "Company").

The CIA has done more war fighting lately precisely because it can change course faster than the supertanker-like Defense Department. CIA officials like to remind people that they put the first "boots on the ground" in Afghanistan after September 11; the Agency's paramilitary operators linked up with Afghan Northern Alliance fighters just a few weeks after the attack. CIA officials also let it be known that they took some of the first shots in the war, firing Hellfire missiles off a Predator unmanned aircraft at a motor convoy thought to be carrying Mullah Omar.[1]

The CIA has to go through most of the same budget hoops as the Defense Department. The difference is that the CIA is not bound by the trifurcated culture and sheer size of the Pentagon (the Defense Department's budget is nearly *one hundred* times larger than the CIA's).[2] If the CIA needs to, it can quickly buy most of the pieces for conducting a good-sized war—troops, aircraft, alliance partners—off the shelf and on contract.

But the CIA could not do most of these operations if the Defense Department did not exist. For instance, one prerequisite for being a CIA paramilitary is, not surprisingly, experience in the U.S. armed

---

services into a single room when necessary. The Joint Staff works for the Chairman, not the Joint Chiefs, and the Chairman does not depend on the Joint Chiefs for his authority. And each of the members of the Joint Chiefs understands that his main job is running his branch of the military.

forces. Johnny Michael Spann, the CIA officer killed in Mazar-e Sharif in November 2001, was a former Marine captain, for example. Similarly, the Predator was originally developed by the Air Force.[3]

The CIA can move people and money fast. It can even operate a "skunk works" to put new hardware in the field faster and with less red tape. But it takes a bigger organization to develop weapons and train soldiers year in, year out. So, when it comes to preparing for war, the Defense Department is the only game in town. As a result, no matter how good an idea for a new weapon or strategy might be, it has to make it through the Defense Department labyrinth.

Suppose, for example, someone had a new idea for attacking computers—say, a drone that could hover over a targeted building and, on cue, send an electronic pulse that zapped all the electronic gear in the house. You would need to convince a CINC that he needed that new capability. He might issue a requirement. You would also need to convince one of the services to fund the project. And you would have to get the project through OSD, which would ask whether the project complies with its priorities and policies.

In addition to all of this, you would have to deal with the military culture, an oil-and-vinegar mix of entrepreneurial creativity combined with conservative tradition. The military culture simultaneously encourages innovation while it discourages new ideas.

In fact, possibly the most remarkable thing about the Defense Department is that, even in an organization largely designed for orthodoxy, people really can make big changes. You just need to be really smart, aggressive, and cagey—and willing to push an idea without pushing so hard that the system spits you out.

Fortunately, there are officers like this, and some of them got together in the early 1980s in an effort that became the foundation of most U.S. information warfare activities today. Not much has ever been officially released about this activity, and you won't see the details here either. The thing is, you cannot talk seriously about information warfare without at least mentioning this project, because so many of the people responsible for establishing U.S. information warfare policies and programs came out of it.

Suffice it to say the objective was to shape the perceptions of Soviet military leaders just before, or during, a nuclear war. The main goal was to keep the Soviets from getting a "go" signal out to their missiles and bombers. The operations involved breaking into Soviet communications networks and fiddling with the traffic going through them. It was a spin-off of another activity: signals intelligence.

Signals intelligence (SIGINT)—intercepting enemy communications—has always been intertwined with information warfare because the first step is usually the same for both: Find a communications link and tap it. Or find a computer and hack it. Once you get in, you can either listen passively or actively insert your own data.

Recall John Boyd and you also see another connection between signals intelligence and information warfare. The two may seem different, but the objective of both is the same: getting ahead of the other guy's decision cycle. You can listen to an opponent's communications, learn his plans, and anticipate his moves so that you get ahead of his decision cycle and can kill him. Or you can jam his communications, deceive him, and make him fall behind your own decision cycle—so that you can kill him.

It's all part of the Perennial Question, and all part of the same package. In fact, the links between signals intelligence and information warfare is so old that you can trace the connections to the Civil War. Communications, signals intelligence, and information warfare all involved the same people.

Armies had used flags for centuries to give soldiers a signal to attack or retreat, or simply to get their attention. But the Civil War was the first time armies used them in a well-organized communications system. It all started when, in 1851, Albert J. Myer, an assistant surgeon from Buffalo, New York, wrote his doctoral dissertation, "A New Sign Language for Deaf Mutes." Myer's new signing system was based on Morse code, which had itself been invented just fourteen years earlier by Samuel Morse when he was developing the telegraph.

The Myer signing system for the deaf consisted of moving one's hand in one direction or another to indicate the equivalent of dots and dashes. After he joined the Army in 1854, Myer combined these ideas—flag signaling, Morse code, and hand signing—to come up with "wigwag," the first modern military messaging system. Using a network of operators perched atop 125-foot towers and equipped with flags, torches, and telescopes, armies could transmit messages over distances of ten to fifteen miles, day or night, at a rate of about three words per minute.

In 1860 Congress authorized the Army to spend $2,000 on signal equipment and to appoint an officer to oversee the new program. The post went to Myer. When the Civil War broke out the following year, some of Myer's men joined the South; the founder of the Confederate Signal Corps, E. Porter Alexander, had even been Myer's student.

Once the war began, the Federals and Confederates both started using flag signals. Since both sides knew each other's methods, it did not take long for both armies to start intercepting their opponent's communications and exploiting them for intelligence. Soon after that, Myer took the idea to the next step—not just intercepting messages, but inserting deceptive messages into the enemy's wigwag network. It was, in effect, one of the first digital network attacks, and this is how it happened:

In April 1863 the Federals and the Confederates were in the middle of a four-year struggle for control over the Shenandoah Valley, the broad stretch of meadows and fields that runs north to south across the middle of Virginia. The valley was not only the main invasion route between Washington and Richmond; it was also the region's breadbasket, and both armies needed it to feed their troops.

The Federals and the Confederates both used cipher wheels to protect their sensitive military traffic. A cipher wheel, the state-of-the-art technology of its day, was a handheld device that looks like a circular slide rule and substitutes one letter on an inner ring with a replacement letter on the outer ring. The alignment of letters on the two rings can be adjusted to vary the match—in effect, creating a "cipher for the day."

But encrypting messages required additional effort, as did translating messages when receiving. So signal operators typically sent routine messages without encryption, or, as we would say now, "in the clear." Operators on both sides also often "chattered" using unencrypted wigwag to pass the time or gossip.°

°Although the terms "code" and "cipher" are often used interchangeably, there is an important difference. A code disguises a message by substituting words for words. Ciphers substitute individual characters for characters through a much more complex process.

A cipher consists of two parts: an algorithm, the formula for scrambling words, letters, or numbers; and a key, the particular settings used by the algorithm to scramble a particular

Although both sides soon broke their opponent's cipher, there was a critical difference in how each treated their success. The Confederates discussed their codebreaking achievement in their routine traffic and chatter. The Federals concealed their breakthrough by ordering their officers not to discuss it in their official encrypted messages, routine traffic, or chatter. Today, a defense intellectual would probably say that the Federals enjoyed "battle space information dominance." That is, the Federals broke the rebels' cipher. They knew that the rebels had broken their old cipher. The Confederates, meanwhile, knew none of this. The Federals thus had the edge.

Myer, based in Washington, proposed a ruse to take advantage of this situation. He composed a bogus message, designed to seem like unofficial unencrypted chatter between Union signal officers. His bogus message implied that the Federals under Brigadier General Daniel Butterfield were moving west to Warrenton. He then sent the message in the clear. Butterfield was actually marching south to Fredericksburg.

Myer knew the Confederates would read the intercepted message, and hoped they would believe they had picked up a juicy piece of intelligence. And sure enough, the Confederates took the bait and ordered, via coded wigwag, Major General J.E.B. Stuart's cavalry to ride west to protect their flank. The Union forces intercepted and decoded the message, so they knew their ruse had worked. This kind of feedback, incidentally, is essential to any good information operation; otherwise you are merely shooting into the dark.

The Federals gave Stuart time to move out. This left a twenty-mile gap in the rebels' lines. The Federals charged through the opening. A week later the Federal forces were still advancing south, and the Confederates were still waiting for the attack on Warrenton.[4]

Myer was later promoted to brigadier general. After the war, he went on to establish the Weather Bureau. Before meteorological satellites came along in the 1960s, weather forecasting depended on local stations across the country pooling data to describe conditions across

---

message. To read an encrypted message, you need both the algorithm and the key. To make the cipher harder to break, you change the key often enough so that it will be hard for anyone to deduce the cipher by intercepting a series of encrypted messages.

large regions. As a result, meteorology and communications have always been closely related, and, indeed, one reason Myer became interested in signaling was because of his interest in weather forecasting. In effect, Myer was the great-grandfather of two of the largest organizations in the federal government: the National Security Agency and the National Oceanographic and Atmospheric Administration.

One hundred years later, when the Defense Department took its first steps in modern information warfare, the technology was different—electronics instead of wigwag—but the basic principles remained the same. The U.S. Navy seems to have led the other services in developing concepts for information warfare in the 1970s and 1980s. This was partly because of the Navy's codebreaking experience in World War II, but it is also the result of how navies work—and the Soviet navy in particular.

Navies have always operated at the ends of long, narrow communications lines. Until shortwave radio came along, ships at sea were for all practical purposes incommunicado. The Soviet navy, however, had an additional problem—communicating with its bases, which were situated at the eastern and western ends of the Eurasian landmass and separated from Moscow by some of the most inhospitable terrain imaginable. As a result, much of the Soviet navy was connected to Moscow by just a limited number of communications links. Because the Soviet navy had so few links, it was easier for U.S. intelligence to understand the layout of its networks and get inside them.

Meanwhile, the U.S. Navy was developing a pressing need to do precisely that. During the Reagan administration, the Navy based its planning on what it called the Maritime Strategy. Like the Reagan national security policy as a whole, the Maritime Strategy was an in-your-face plan designed to stress the Soviets to the limit. It helped that Reagan's Secretary of the Navy, John Lehman, was himself a young, in-your-face political player eager to make his mark. When he was not in his Pentagon office, Lehman was likely to be flying in the right-hand seat of an A-6 bomber as a reserve officer. Lehman was a tough competitor for both the Soviet Union and the other services and was determined in any war to take the U.S. Navy right to the Soviets' doorstep.

The centerpiece of the Maritime Strategy was a plan to destroy the Soviet fleet early in a war. The most important targets were Soviet missile-launching submarines. To get them, the Navy—aircraft carriers on the surface, submarines below—would fight its way into the Barents Sea, the icy waters above the Arctic Circle off the coast of the Kola Peninsula. An ambitious plan, to say the least, but it was just the kind of challenge ambitious Navy officers would leap at.

"It was a great time to be in the Navy. We had the only plausible approach of all the services," recalls Greg Blackburn, who had a hand in the early planning, and still sounds like he's ready to pitch the idea in a Pentagon briefing. "We had the concept of operations, identified the targets—we had the whole story."

For the Maritime Strategy to work, the Navy needed detailed, precise intelligence on the communications from Soviet nuclear forces and, if possible, ways of interfering with them. "The Navy's thinking about how to go after the target was more plausible," according to Blackburn. "Once you get on land, the problem is harder. We had a tractable problem and good ideas."[5]

One of the operations in which the Navy cracked into Soviet communications networks that has come to light was a program called IVY BELLS. The Navy tapped a cable running hundreds of miles under the Sea of Okhotsk that connected the Soviet's Pacific Fleet submarine base on the Kamchatka peninsula to the Siberian mainland.

IVY BELLS ran from the mid-1970s until 1981, when Ronald William Pelton, an employee at the National Security Agency, sold the secret to the Soviets for $35,000. The "pod" that the Navy divers attached to tap the cable was about twenty feet long and looked something like a telephone-switching terminal. The pod is now on display at a museum in Moscow run by the SVR, successor of the KGB. Pelton is on display in federal prison.[6]

If you believe the press reports, had Pelton not betrayed his country, the best was yet to come. Rear Admiral John Butts, the Director of Naval Intelligence, proposed tapping a cable in the Barents Sea that ran to the Soviet Union's main submarine base. Butts wanted to run a cable from the tap to Greenland, and then send the intercepted traffic to the United States, where analysts could listen to conversations as they occurred. If the analysts heard something like a "go" signal, they could send an alert to U.S. forces.

Angelo Codevilla, a staffer who reviewed the plan for the Senate

Intelligence Committee, described the project. "If you can get a line onto one of his lines without his knowing it, you've got it," said Codevilla. "They could lift up their cable and inspect it; it would be clean as a whistle. They would lay it down again, they would lay it right back down on our listening device." Once you accomplished that, Codevilla said, "you simply know everything he knows. You are inside his circle of decision." Inside the decision cycle—John Boyd would have been pleased.[7]

IVY BELLS was just one program that got into Soviet communications networks. Throughout the Cold War, the United States flew aircraft right up to the Soviet border to listen to military communications, intercepted shortwave and satellite communications from the ground, picked off microwave links from space, and tapped phone lines.[8]

It is not much harder to put messages into a successfully tapped network than it is to take them out, just as Albert Myer did in 1863. U.S. commanders, though, needed more than a neat way of screwing around with Soviet communications. They needed reliability and predictable effects. So, as these projects multiplied, the Defense Department had to figure out how to turn these individual penetrations into a single, integrated system—that is, a useful weapon.

The answer came as a byproduct of an art form known as critical node analysis. It was part of the targeting process for strategic nuclear war. During the 1960s the Strategic Air Command began to study Soviet communications networks more seriously. Ronald Knecht worked on the project when he was an Air Force officer. "There were some findings in the late 1970s," Knecht says, "that there were vulnerabilities in their networks."[9]

More recently, Internet mapping serves roughly the same function. The question was whether anyone could fashion this knowledge from critical node analysis into a useful weapon. In autumn 1983 the Pentagon created a special group to study the problem. The group was called, aptly enough, the Joint Special Studies Group (JSSG) and was based at the National Security Agency.[10] It drew on critical node analyses, knowing that once you find the right node, you can intercept messages, jam the system at critical moments, or inject your own signals into the network.

These kinds of activities—essentially, military hacking—became known as special technical operations (STOs).[11] To link all of the STOs together, the JSSG hired Hughes to develop a central command and control network linking the Joint Staff, unified commands, and intelligence agencies with the STOs. This network evolved into the current Planning and Decision Aid System, commonly known as PDAS.[12]

When the JSSG finally got its system put together, it found that it could model the technical characteristics of Soviet networks—the communication lines, the devices at each node, and so on—but the human operators were too unpredictable. In the end the JSSG had a system that would complicate a Soviet commander's communications but couldn't guarantee that he would be unable to get a "go" signal out. The system wasn't a useful weapon. The Pentagon pulled the plug, the Cold War ended, the Soviet Union broke apart, and the United States and Russia took their forces off split-second alert.

Intelligence officials had qualms about the project in any case—again, it was the Perennial Question: *Deny, deceive, destroy, or exploit?* The warfighters wanted to keep the enemy from using its communications systems. The intelligence gatherers wanted to listen in, and were rightly queasy about anyone tipping off their target by jamming or spoofing it.

But the JSSG had showed how to translate the ideas articulated by Rona and Boyd into reality. The essentials were there: breaking into an information network, getting into position to disrupt the enemy's decision cycle, and then leveraging that opportunity to win a battle.

Chapter 7

THE WAR NEXT TIME

On a sunny day on June 8, 1991, 800,000 people lined the streets in Washington to watch the parade celebrating the U.S.-led victory in Desert Storm. An F-117 stealth fighter opened the event in a surprise flyover, racing down the length of the Mall from the Capitol to the Lincoln Memorial. The capital had not seen such a celebration since Dwight Eisenhower returned from Europe after the defeat of Germany.

Even as the tanks and missile launchers were lining up for the parade, back at the Pentagon people were passing around a Tom Rona essay, "The War Next Time." By now Rona had left the government. He had written a wry piece that took the form of a memo offering advice to Saddam on how to do better next time. Rona, with a sense of humor, had even gone to the trouble of printing his essay so that it looked exactly like an op-ed from the pages of *The Wall Street Journal*.

In fact, the piece had never appeared in the paper. "I just decided to cut out the middleman," Rona joked. In the future, he wrote, don't give the United States free use of its satellite communications. Don't fight the war in the middle of the desert, where smart weapons work so well. Don't fight from fixed positions; trenches and dug-in tanks are easy to spot from space. And hire a better public relations firm so U.S. leaders have a harder time rallying public support.

The high-tech American weapons had amazed everyone. Rona was trying to remind them that this information edge was fragile. If Iraq had just taken some simple countermeasures, that advantage could have quickly disappeared. The United States did not win the war just because we were so smart; it was largely because Saddam was so stupid. Counting on an enemy to be stupid was risky.

Pentagon officials like Duane Andrews shared Rona's concerns. They fully appreciated how kind technology had been to the United States, and worried about whether it would be so kind in the future. In fact, they began to worry that adversaries would use our dependence on high tech against us.

Andrews was Assistant Secretary of Defense for Command, Control, Communications, and Intelligence. In 1989 Andrews, a former Air Force officer, had been a staffer for the House Permanent Select Committee on Intelligence for twelve years. He might have stayed there had it not been for Paul Weyrich, president of the Free Congress Research and Education Foundation, a conservative activist group.

After George Bush won the 1988 election, he nominated his fellow Texan, Senator John Tower, to be Secretary of Defense. Bush and Tower were close, having been Texas Republicans together back in the days when such a thing hardly even existed. But Tower had never been widely popular among his Senate colleagues. His nomination inched along until Weyrich let it slip in one hearing that he had sometimes "encountered the nominee in a condition—a lack of sobriety—as well as with women to whom he was not married."

Perhaps the conservative Tower had crossed Weyrich in the past. Perhaps the conservative Tower just was not conservative enough. As Weyrich says, "The Free Congress Foundation is politically conservative, but it is more than that: it is also culturally conservative."[1]

It really did not matter. Opponents had the cover they needed, and two months later Tower's nomination went down in a party-line vote. Bush quickly picked Dick Cheney, the Wyoming Congressman, as a replacement. Cheney had been a member of the House Intelligence Committee, knew Andrews well, and soon the staffer was moving from the committee's office under the Capitol dome across the river to the Pentagon.

By that time, Ron Knecht was a colonel with twenty-nine years in the Air Force, wondering what he would do after retirement. Andrews called Knecht and asked if he would accept a political appointment.

Andrews had gotten to know Knecht over the years as a straight shooter who would tell him whether an Air Force program really worked.

After working on special access programs for more than a decade, Knecht had a pet gripe: secrecy. Specifically, when it came to weapons for information warfare, he thought there was too damn much of it. "I'm coming with an agenda," Knecht recalls warning Andrews. "We've got to do something about information warfare and get it out of the compartmented world."

The problem with making any weapon too secret is that the troops never get a chance to become familiar with it. Lack of familiarity equals uncertainty, and military commanders *hate* uncertainty. Swashbuckling, seat-of-the-pants generals exist mainly in movies. In the real world, all commanders try to reduce uncertainty to the lowest level possible. Uncertainty means you don't know where the enemy is or what he plans to do. Uncertainty means you are not sure what your own forces can do. Uncertainty gets you killed.

Uncertainty includes earnest technoids who show up on your doorstep just before the fighting begins, toting gee-whiz electronic stuff from behind green doors secured with combination locks. Knecht believed in the power of information warfare, but he knew that commanders would use it only if it was a familiar tool that they had already worked into their battle plans. He reported for work in January 1990.

Knecht's ally was Paul Strassmann, a retired executive from the Xerox Corporation. Strassmann had a colorful history of his own. He was, as he put it, "the son of a rich Jew" growing up in Slovakia when World War II began. Orphaned, he joined a resistance group and fought as a fifteen-year-old guerrilla against the Nazis. "I was exposed in the underground to what the Communists were like," he recalls. "Especially the crooks."

Like Tom Rona, Strassmann made his way west after the war, first through France, then Britain, and finally reached the United States in 1948. The guerrilla fighter had wanted to be an engineer since he was six and had heard about the Cooper Union, the full-scholarship college in New York City. He got in, graduated, and got a job as a civil engineer working on the New Jersey Turnpike.

The turnpike authority was having problems with traffic jams at the tollbooths. Strassmann's boss asked him to figure out a solution. "I was told to use a computer to explore the situation. They let me play with it day and night. I had the turnpike's punch-card toll receipts, and I fed them into the computer and analyzed traffic patterns. It turned out that the problem was the scheduling of the toll collectors."[2]

Strassman reworked the schedules and was a hero. It was his introduction to computers. By 1969 he had joined Xerox, and eventually he was in charge of all of its corporate computer operations. He had been retired for almost five years when Donald Atwood, the Deputy Secretary of Defense, asked him in 1990 to join the Defense Department as a consultant. Atwood was a former GM executive and was responsible for running the nuts and bolts of the department while Dick Cheney dealt with national policy and the rest of the outside world. Atwood needed someone to fix the Defense Department's computer operations, which had grown like kudzu from a hodgepodge of systems over the years.

This project started Strassmann thinking more about computers as targets. Part of the problem of making a computer system reliable is making sure that the operators cannot sabotage it. "I was the computer guy, so I continually brought up the issue of how someone can tinker with these systems," Strassmann says. "The insider mole was the worst threat."

Strassmann had landed in the middle of some of the early thinking about cyber security and defensive information warfare—the flip side of offensive information warfare, what Knecht was working on. Strassmann also went to work for Andrews.

After Desert Storm, there were some tales, mainly false, about how the American forces hacked into Iraqi computer systems. According to one, the U.S. had somehow managed to sell Iraq printers that had been rigged to infect computer networks with viruses. Or maybe it was the French who planted the printers. In any case, the story was a hoax. But there were other true, if less exotic, operations.

"In Desert Storm, for the first time, we employed the technologies of what would be called offensive information warfare," Andrews recalls.[3] Most of the operations were aimed at crippling Iraq's military networks. Of course, armies had jammed communications and spoofed air defenses for decades. This, however, was the first air campaign specifically designed to bring down a country's communications system in a carefully phased plan. On the other hand, Andrews believed that we might easily come out on the losing side of an information war next time. "In Desert Storm, we began to see that a relatively unsophisticated enemy could do significant harm to us."

"The first thing we did was go to Colin Powell," Andrews says. Powell was then Chairman of the Joint Chiefs. "We talked about the problems of Desert Storm, and Powell gave us his support." The immediate objective was just to generate a piece of paper—a Defense Department policy directive. "The whole thing got started by Ron Knecht," recalls Strassmann. "The term 'information warfare' had to be established as a hook. Ron understood the directive would serve as that." Robert Carpenter, a captain who had been involved in the Navy's signals intelligence operations for years, was brought on board to do the drafting.

The directive would tell the services—Army, Navy, and Air Force—that the Defense Department corporate office believed information warfare was a legitimate function, and it was okay to develop weapons and train fighters for it. Since Powell had signaled his approval, the services would know that the uniformed leadership and civilian leadership agreed that information warfare capabilities were needed.

All of this would, in principle, clear the way for organizations and offices inside the massive Defense Department bureaucracy to propose information warfare programs. Probably only a few people at the top really had an understanding of the nitty-gritty that was underway inside the services; there are so many programs and projects—literally thousands of budget lines—that few people outside the Comptroller's office can keep track. Classification also keeps people from knowing what magic widget is being developed around the corner. But that was okay.

In one limited sense, the Defense Department—with the fall of the Soviet Union, now the largest centrally planned economy in the world—resembles the free market. Just as individual consumers have little influence over what kinds of cars Ford or Chevrolet build, OSD and the Joint Staff could not give detailed instructions to the services about how to prepare for information warfare. All they could do was signal that they wanted to see proposals, just as someone shopping for a car can show up at a dealer with an open wallet.

By September 1991, though, Duane Andrews had a new problem: George Bush's prospects for re-election began to tank, thanks to the slowing economy and Ross Perot, who was splitting the Western state vote. Andrews, Knecht, Carpenter, and everyone else working on the new policy directive were now racing against the clock. It was not just that the new administration might have its own ideas about cyberwar. The reality is that in the Pentagon all new ideas freeze in place for a year or more whenever a new President is elected. Everyone waits for the new team to get through the appointment and confirmation process. Short of a war, it is hard to get anything new accomplished.

All of this explains why Department of Defense Directive TS3600.1, *Information Warfare,* was classified Top Secret. It was a matter of expedience. Andrews and everyone else who was a Bush administration appointee needed a policy before January 20, 1993.

Andrews and Knecht understood that pushing their policy directive through the Pentagon would take even longer if they tried to get every agency involved in information warfare to declassify their materials. It was easier to simply classify the entire directive Top Secret. They managed to get through the hoops, and on December 21, 1992, with exactly a month to spare, Defense Secretary Cheney approved the directive.

Ironically, the fact that TS3600.1 was Top Secret meant that Ron Knecht's policy had, in at least one respect, exactly the opposite effect from what he had intended. Knecht believed that information warfare had become too tightly compartmented, too secretive. But by classifying the directive Top Secret, the Defense Department was formally pulling information operations, including some that had been routinely discussed in open channels over the years, even more tightly into the secret world.

Even so, just seventeen years after Tom Rona coined the term, "information war" was no longer just an idea. It was now the policy of the U.S. Department of Defense. During the next ten years the Defense Department issued two revisions of the directive, lowering its classification to "Secret" in 1996 and finally issuing an unclassified draft in 2002.

The new directive was a big step, and it is hard to overestimate the significance of this particular piece of paper. The impact of TS3600.1 on how the U.S. military deals with information was in the same league as, say, the impact of the breakup of the Bell System on commercial telecommunications. It changed the rules of the game.

The Defense Department had used information technology for more than a century. Now, though, the Secretary of Defense had said that there was an entire mission area based on attacking and protecting information systems, and the Chairman of the Joint Chiefs had endorsed it.

So officers could plan careers knowing that information warfare had a place in the department's thinking. The services could propose programs to fight a cyberwar without having to first make the case that military commanders would use them. Commanders knew that information warfare was something they had to address in developing their war plans. Information warfare, to put it simply, had arrived.

Chapter 8

CYBERWAR IN THE DESERT

"I've gained a new appreciation for stock car racing," David Ronfeldt announced. He and I were having lunch in Chez Jay. It was dark inside—even on a sunny spring afternoon—and noisy, but after a while, most of the noontime crowd began to drift out.

David and I had both been lamenting the decline of Formula One, the classic international auto racing series. Colorful racers like Stirling Moss and Jackie Stewart were a thing of the past. Grand Prix drivers had all turned into plain vanilla multimillionaires with coiffed hair and personal corporations run by high-priced business managers out of Zurich via Cayman Islands tax shelters.

Even worse, the races had become boring. Over the years, all the teams had analyzed each other's designs to the nth degree. They had, in effect, managed to perfect the design of the Formula One racer. So now everyone was evenly matched. Even worse, all the classic racetracks—Spa, Nurburgring, Monza—had been shortened and swathed with double-height steel guardrails so that each looked like an identical rain gutter running through an amusement park.

Nowadays once the field left the starting grid, most Formula One races turned into a parade. Hardly anyone passed a car during the race. We mused that perhaps Europeans might not care how often, or even whether, cars passed each other. After all, these were people who went bonkers over soccer, where games seem to run for hours—no, days, even weeks—without anyone scoring.

At that point David and I began to wonder whether there was a double entendre in our European friends' inability to score, but I still could not fathom why David would like stock car racing. At least Formula One cars drive on winding road courses. Stock cars run on high-banked ovals, so stock car races seemed even more boring.

"There's more information content in stock cars," David went on, explaining the subtleties of competition in NASCAR, the National Association for Stock Car Auto Racing.

Like Formula One cars, stock cars have evolved into a species of androgynous equals. In fact, underneath the skin, most stock cars are virtually identical. So none of the top competitors have a significant speed advantage. The only way one driver can pass another is to "draft"—use a car ahead of him to break the wind so he gets an extra one or two miles per hour—and then slingshot around him.

By the time David's soup arrived, he was well into making his argument that it was precisely this mechanical equality that makes a stock car race an information-heavy game of strategy. Because stock cars, unlike Formula One racers, race on high-banked ovals, they basically run flat out—that's where the expression "put the pedal to the metal" came from. To stay in the hunt, you need to be part of a line of cars drafting each other. Otherwise you fall off the pace. If you stop to change tires or refuel, you must make sure you can get into a new draft line when you return to the track.

All NASCAR teams have radios in their cars, so the pit crews talk to their drivers throughout the race. In the pits, teams talk to each other by radio or cell phone to make deals with other teams to time their stops for mutual advantage. You can always break a deal if a better opportunity comes along (like a faster draft line), assuming you don't get outmaneuvered first. And stock car teams have even been known to spoof their competitors with bogus messages, or fuzz up another team's communications by having someone transmit on their frequency.

The trick to winning the race is to jockey for position so that you are near the front of the pack during the last four or five laps. The best position is to be in second or third place going into the last turn so you can slingshot ahead at the last minute, crossing the finish line ahead of the car that had been leading. It was all a contest of bargain and maneuver.

Social dynamics combined with information technology, David argued, to make stock car racing the perfect sport.

I was impressed. It was elegant. It was logical. It was counterintuitive. Hell, it was probably even true. In short, it was a classic RAND argument, possibly even more significant than Marshall and Wohlstetter's discovery of the asymmetric threat. And, although it seemed to be nothing more than an observation about redneck good ole boys chasing each other around a circle at nearly 200 miles per hour, it really offered insight into more important problems. David published the results.[1]

Information technology was changing stock car racing in many of the same ways it was changing warfare. Hardware had become a great equalizer; just as all the racers had comparable equipment, so did most warriors. To compete in either world, you needed networked telecommunications and organizational agility. The winner was usually the team that could collect, process, and apply information most effectively.

John Arquilla was a former Marine who had just finished his doctoral degree at Stanford before coming to RAND in the early 1990s. He and some others in his group had worked for Central Command during the war and had been involved in some of the planning behind the Left Hook—the flanking maneuver Coalition forces used to win the Gulf War. The Coalition forces had taken a "western excursion" into the desert and made a wide turn to bypass and surround the Iraqi forces, which were mainly bunched up near the Gulf.

The Left Hook was a classic flanking maneuver, and flanking maneuvers are as old as armies. Cavalry depended on the fact that horses run faster than soldiers to "turn the corner" of an enemy formation. Later tanks and mechanized infantry did the same thing, except even faster, and in the 1970s the U.S. Army made the flanking maneuver three-dimensional by using helicopters in what it called its AirLand doctrine.

Navies also had their own form of maneuver that was an analogue to an army's flanking move. It was called "crossing the T", or navigating so that your battle line cut across in front of the opponent's line. This allowed you to let loose a full broadside against your adversary's bows, while he could not return fire because most of his guns pointed to the sides.

In all of these cases, the idea is the same: Military formations are designed to move and concentrate their fire in one direction. So if you can move faster or with greater agility than your opponent, you can get into an advantageous position where you can fire at him more effectively than he can fire at you. The alternative to outflanking your opponent is to outgun him with heavier and longer-ranged artillery, or throw enough bodies at him that he runs out of ammunition before you run out of bodies. For centuries these were the two basic options: Overwhelm your opponent with mass and firepower, or outmaneuver him with speed and agility.

The Left Hook had been conceived as a traditional flanking maneuver, and even then it was not the Coalition's first choice for a strategy. After Iraq invaded Kuwait in August 1990, the main goal of American leaders, military and civilian alike, was simply to stabilize the situation and make sure the Iraqis did not roll on down into Saudi Arabia. There was nothing between the Iraqi army and the Saudi capital of Riyadh but 275 miles of empty desert. Saudi forces were so weak that the CIA thought the Iraqis could cover the distance in three days.[2] At the time, the United States had no troops in the region; the Saudis, protectors of Islam's holy sites, had never allowed infidels to base troops in their land. Simply rushing the first troops to the scene required four months, and it was only in October that American leaders began to think about an "offensive option."

It was only logical that the first plans General H. Norman Schwarzkopf offered began by proposing that the Coalition forces would move from their defensive position to attack the Iraqis head-on. Nobody really liked this option. Everyone knew that Iran had lost thousands of soldiers using exactly that strategy against Iraq.

Defense Secretary Dick Cheney derided the plan as "high diddle diddle up-the-middle" and began to look for alternatives. As it happened, Henry Rowen, his Assistant Secretary of Defense for International Security Affairs, had been reading Sir John Bagot Glubb's *A Short History of the Arab Peoples*. Glubb was a British army officer who had founded the Arab Legion in Transjordan. In his memoirs, he told how his troops had maneuvered in the deserts of western Iraq during World War II. Rowen thought American troops could do the same to bypass the Iraqi defenses. Soon Cheney had the Joint Staff in the Pentagon fleshing out an alternative plan.

CINCs do not care for anyone—let alone Washington, and much less civilians—to tell them how to run a war. Schwarzkopf, who had a famous temper in any case, naturally exploded when he found out. But in fact, Schwarzkopf did not really like the idea of a head-on attack either. The only problem was that he believed any alternative, especially a flanking maneuver, would stretch his forces too thin.

After everyone thrashed out the options, they gradually converged on the plan that became known as the Left Hook. First the Americans would send another corps—145,000 troops—to Saudi Arabia (mainly by pulling forces from Europe). Then, assuming negotiations failed to get Iraq out of Kuwait, the Coalition would undertake an air war to beat down the Iraqis. Then, assuming that Iraq still would not budge, the Coalition would follow up with a flanking maneuver around the western end of the Iraqi defenses.[3]

Only later did analysts appreciate how this plan enabled the Coalition forces to enjoy a new advantage, an advantage that distinguished the Left Hook from flanking maneuvers used in earlier wars. The Coalition forces had information superiority and took full advantage of it.

Early on the morning of January 17, 1991, a group of eight Apache attack helicopters, Task Force Normandy, slipped across the border separating Saudi Arabia and Iraq. A Pave Low helicopter, originally designed to support special operations forces, guided the Apaches.

The group flew fifty feet over the desert floor at speeds of up to 120 miles per hour. Traveling 700 miles in the darkness, they reached within twelve kilometers of two radar sites west of Baghdad. On signal, the group destroyed the radar sites with a round of missiles and machine-gun fire, putting a hole in Iraq's warning network.[4]

Then a round of cruise missiles, timed to arrive over Baghdad a few minutes later, dropped spools of graphite fibers over transformers and switching yards outside the city's power stations. These fibers shorted out circuits, causing the power stations to shut down and putting Baghdad—and most Iraqi political and military headquarters—in darkness. After that, attack aircraft, rushing through the hole in the radar network, dropped smart bombs on Iraqi command and control bunkers that had been identified during the preceding months.

Within a few hours the Iraqi government's central nervous system was stunned, severed, or destroyed. During the next four weeks the

Coalition cut most, but not all, of Iraq's communications lines. When the Coalition armies then disappeared into the western desert, not only did the Iraqis not know where they were, they were confounded by bogus information planted by Coalition forces coming through the surviving communications lines.

At that point speed, usually critical to turning the corner in a traditional flanking maneuver, was not the most important factor. Even if they had not been fast, the Coalition forces could have outflanked the Iraqis, simply because the Iraqis could not see them, had no idea where they were, and did not know where they might reappear. Coalition commanders, on the other hand, knew where their forces were, because each unit knew its own location and was hooked into a secure, reliable communications network.[5]

Desert Storm was the first war in which an army came close to operating a true networked communications system. The Coalition had layer upon layer of links: local radio, military satellite, commercial satellite. The line between local communications and global communications blurred. It did not always work smoothly, and a lot of the network was jury-rigged, but the glimpse into the future was clear.[6]

Even Arquilla was surprised by just how successful the move had been. He thought about it. About a month after the end of the war, he stuck his head into Ronfeldt's office and announced, "I have just one word for you: cyberwar."

Arquilla and Ronfeldt began to imagine, what if instead of sending a single large wave of forces around an enemy's flank in a massive, floating, networked Left Hook, we used *all* forces like that? In other words, suppose every unit operated like a separate detachment, linked via communications to everyone else. They could then swarm on their victim at an appointed time, much like bees.

This seems to have been the first time anyone had come up with the idea of networked armies. It seemed brilliant and inevitable all at the same time. Just as cheap computers and time-sharing drove Tom Rona to write about information war, another new communications technology and the system that made it possible—the Internet—inspired Arquilla and Ronfeldt to propose the idea of cyberwar.

The Internet traces its roots back to 1964, when the Defense Department's Advanced Research Projects Agency (ARPA) began experimenting with a new kind of computer network based on "digital packet messaging." As one might expect, this new technology was made possible by two ideas: digits and packets.

The original way to transmit a voice message was by sending analog messages through a line, the way telegraphs, telephones, and wireless had always worked. There was a continuous link—a copper wire, glass cable, or radio frequency—between two parties. To send a message, one modulates the energy passing through the link.

In effect, an analog transmitting device encodes a voice or text message into a series of electronic ebbs and flows in the electromagnetic energy flowing through a line or over a radio frequency. A receiver picks up the transmission at the destination, and converts the electromagnetic ebbs and flows back into the original message, just as though you were speaking into a long hose, with someone else listening on the other end; the message is an electromagnetic analogue of the message itself.

This is a simple, straightforward approach that served everyone well for over 100 years, from the time Alexander Graham Bell called his assistant in the first telephone message: "Watson, come here; I need you." The main drawback of this approach is that analog communications, by its nature, requires a continuous link between sender and recipient.

In digital communications, there is an additional step that adds some complications, but offers ways of getting around this constraint, as well as other advantages. Instead of sending an analogue of a message, a digital transmitter first encodes the message. In effect, it writes a description of the message consisting of ones and zeros, and then transmits that description.

Once you go to the trouble of putting a message into standardized digital form it is simple to recalculate the ones and zeros with a set of instructions—that is, a computer program. Digital messages are easier to amplify, verify, duplicate, and manipulate in all kinds of ways so as to cram more messages through a link.°

Several companies, like Bell Labs and Northern Electric, began ex-

° Strictly speaking, one might want to consider the early telegraph as digital technology, since it worked by sending dots and dashes. However, these telegraphs had no way to manipulate the binary data.

perimenting with digital communications technology in World War II. (These are now Lucent and Nortel, respectively, following two decades of mergers, divestitures, and transformations.) The main obstacle was inventing a really cheap, efficient switch that could manipulate electronic digits, which is exactly the task that transistors and, later, microprocessor chips perform. Communications began to go digital in the 1970s, and digital communications really began to take off in the 1980s, along with everything else that depended on computer chips.[7]

You can send a digital message through a dedicated link, just like an analog message. But once you convert to a digital format, it then becomes possible to add another step that offers yet more advantages: packet messaging.

Instead of using a single, continuous line or a radio frequency between the sender and the recipient, in packet messaging one first chops the digital message into uniform data packets. Each packet is tagged with a header containing instructions that tells where it is supposed to go. A computer then ejects the packets into a network, and each piece is directed to its destination by any computer that it happens to pass through along the way. Once all the packets arrive at their destination, the recipient reassembles the message.[8]

This may seem like a way to communicate that is even more roundabout than going digital, but it offers some advantages. A message can reach its recipient as long as there is *any* continuous chain from point A to point B. The message does not depend on any single, specific link. So, even if several lines or computers in a network were destroyed, the network can still function. This was the feature that had originally caught the interest of the Defense Department—the potential to build a communications system that could survive a nuclear war.

An even more important benefit that did not become clear until later was that it was easy for additional computers to join the network. As long as you complied with the network's software standards—"protocols"—you were in business. You could dump a message into the network, and it would find its way to its intended recipient.

British technicians at the National Physical Laboratory did the first experiments with a digital packet message network in 1968, and the next year ARPA funded a larger-scale test project. This network,

ARPANET, consisted of four nodes and first went into operation on October 29, 1969, when Leonard Kleinrock and Charley Kline tried to send a message from their computer at UCLA to a computer at Stanford. The message: "LOGON." Kline got as far as "LO." Then the system crashed.[9]

Kleinrock and Kline got their message through successfully about an hour later. They thus set the pattern for millions of future Internet users: click, receive error message, reboot, and try, try again.

By the next year there were nine nodes on the network. This grew to fifteen in 1971 and thirty-seven by 1972. The network grew exponentially from there, first among the defense and scientific communities. The National Science Foundation began funding the system in 1984, and five years later ARPA officially phased out the Defense Department's participation, leaving the civilian Internet that we know today. In the early 1990s businesses and the general public began to log on, and despite the dot-com boom and bust of that decade, the Internet has since become a fixture of modern life.

Decentralized, digital communications systems were a key technology that made network warfare possible. Until then, armies were rigidly bound by the natural geography. A commander would lay his "axis of attack" along major roads or across open fields so troops could quickly penetrate into an enemy's rear once they broke through his front line. The "axis of attack" also kept an army's formation intact, and this made communication easier.

Staying in formation is less important for maintaining communications when you are using a digital network, and so geography also becomes less important. Indeed, instead of geography defining the military formation, the communications network defines the shape and action of a military organization. You deploy your forces along the network.

Ronfeldt was a sucker for anything with the word *cyber* in it, so he and Arquilla began to write a series of papers on the topic.[10] As networked communications became available, organizations—Zapatistas in Mexico, international organized crime in Russia, militant environmentalists in Seattle—would adapt to use it effectively. Businesses would develop ways to use networks to make money; organized crime would develop ways to take money; political groups would use them to take power; armies would use them to make war. Those that did not would be left behind in the best Darwinian fashion.

These four ideas define the most important features of modern warfare. Marshall and Wohlstetter introduced the idea of the asymmetric threat. Rona captured the essential fact of how information technology changed the nature of war. Boyd showed how beating an opponent's decision cycle determined the winner in combat. And Arquilla and Ronfeldt explained how interconnected, digital communications were changing how to organize a military force.

Taken together, these ideas do not simply accelerate traditional methods of war the way mechanization did. Rather, they upset basic assumptions about how to plan and fight wars. They also upset the relationship between strong nations and weak nations. As September 11 demonstrated, even the weakest, most backward nation in the world can now organize a credible strike against almost any other nation, almost anywhere in the world. Indeed, with these four ideas, organizations that are not even national governments can now play on the same field as world superpowers.

As the theories and technology of information-driven warfare have developed since Desert Storm, warfare has changed. The ability to maneuver quickly and concentrate firepower has become less crucial. The new capabilities that decide who wins are:

- the ability to pick off your adversary from a distance with a single shot;
- the ability to maintain a stealthy network of forces so that, if a target becomes vulnerable, you can swarm around it and destroy it at close range;
- the ability to control information so that you can complete your decision cycle before the enemy completes his.

Today stealth is essential to survival. The first strike is now more advantageous than ever before. The agility of a network is more important than pure firepower. And all these tactics are scalable. They apply to both large military forces and small. They apply to rich countries and poor. That is the new face of war.

Chapter 9

ZAPPING

Fighting networks have two attack modes. They can zap or they can swarm. The difference is whether the information technology is in the weapon or the organization. In zapping—let's use the term military aficionados are calling it, "precision strike"—a fighter in one network uses a high-accuracy weapon to pick off some target, often one that unknowingly exposes itself. In swarming, fighters in a network appear out of the ether to swoop in and kill an unsuspecting target en masse.

The two are related and even overlap in some respects. But the success of both depends almost entirely on information technology and, even more important, having superior information.

The idea of leapfrogging the front and striking directly at the enemy's rear is an old one. Airpower theorists like Giulio Douhet and Billy Mitchell had this kind of leapfrogging in mind when they promoted strategic bombing in the 1920s. Strategic bombing came with the arrival of airplanes, and because military strategists began to appreciate how much modern armies depended on an industrial base for support.

Strategic bombing doctrine was also a reaction to World War I and the horrible stalemate of trench warfare. In the 1920s and 1930s trench warfare had the same kind of influence on military thinking and public opinion that the Vietnam war had in the 1970s and 1980s. No one ever wanted to repeat the experience. Strategic bombing promised an alternative.

Precision strike resembles strategic bombing in that it often involves long-range bombers and missiles. But there are important differences.

The key idea behind strategic bombing can be summarized as "destroy a nation's morale and means of production." It was an idea suited to an era of Fabian Socialism, Communism, and the New Deal, concepts of the state based on nationalism and centralized economies. The key idea behind precision strike, on the other hand, is "one bomb destroys one target." The target could be a factory 100 miles behind enemy lines, but then again it could be a tank or an aircraft just ten miles away. Precision strike is simply how you attack an opponent from afar in an era when information technology makes weapons incredibly accurate.

The technologies enabling strategic bombing were long-range aircraft and missiles—Industrial Age technologies for leapfrogging the front. The technologies behind precision strike are target detection and guidance systems—Information Age technologies. Leapfrogging the front is not an issue, because there is no front. Combatants, and everything else, are intermingled.

The point is simply that now you can destroy *any* target with a single shot if you know where it is. To be sure, sometimes you are bound to miss, sometimes you are bound to hit the wrong target. But long-range weapons with a high probability of kill do change the nature of warfare. Combat becomes a process of beating the enemy's efforts at concealment, or waiting for him to make a mistake and expose himself. When this happens, you strike, and more often than not, you kill. It makes beating your opponent to the end of the OODA loop more important than ever, which is to say, it makes winning the information war more important than ever.

The invention of a sure-shot weapon that never misses its target is an ancient pursuit. Just about every culture has its legendary marksman—Switzerland has William Tell, we have Davy Crockett. But no one really considered *guiding* a weapon to its target until about 150 years ago.

The oldest artillery dates back to the ancient Greeks and Romans, who used catapults powered by springy cords—in effect, large cross-bows—that could throw a one-pound rock about the length of a football field. Around the twelfth century, the French introduced the trebuchet, a much bigger catapult powered by counterweights. A good trebuchet could hurl a 200- to 400-pound stone 500 yards or more, or equal masses of dung, dead horses, and unpersuasive emissaries approximately the same distance.[1]

Alas, trebuchets were so big that you had to build them where you needed them, and that took several weeks. Once explorers brought gunpowder from China to Europe, armies could make transportable cannon, and as metallurgy improved in the late 1800s, so did the range of artillery. Today's artillery has a range of about twenty or thirty miles.

But no matter how far a catapult, trebuchet, or gun could shoot, they all worked the same way—ballistics. After a shell left the gun barrel, you could not correct its flight. It landed just about exactly where Newton said it would. Feedback, the basis of any guidance system, is the electromechanical equivalent of the OODA loop. And the more efficient the feedback, the more likely the weapon is to hit its target. The feedback loop for artillery was, in effect, locked in—it was equal to the time it takes for a shell to reach its target.

Artillery designers continued to make improvements, most recently by adding information technology. In Desert Storm, American tanks had a huge advantage over the Iraqis in range and accuracy, thanks to computer aiming systems. Guns on American tanks compensated for wind, terrain, temperature, and even that minute amount of droop that occurs in a twenty-foot-long gun barrel.

As a result, a single shot from an M-1A Abrams tank has about a 90 percent probability of hitting its target from a distance of two miles. Also, because the aiming process is computerized, American tanks can fire accurately even while racing along at forty or fifty miles per hour. Iraq used Soviet-built tanks, which used older technology. By the time the Iraqis moved within range, spotted their targets, and returned fire, they were dead.

The Abrams's aiming system can calculate the effect of more factors, so its round is on a more accurate trajectory when it leaves the barrel. Even so, after that, the round is committed. If the aim is off, if some unaccounted factor upsets the round in flight, or if the target simply moves, it's tough luck. No correction is possible until the next round is fired.

Right now, the Abrams's 90 percent kill probability is about the state of the art of traditional ballistic artillery technology. To improve your ability to hit a target, you need a faster feedback loop. This requires an active guidance system. Indeed, one definition of a guided weapon is any weapon that can correct or adjust its course in less than the time a round requires to reach its target. The first such weapon was the self-propelled torpedo, developed in the late eighteenth century.

Today we think of a torpedo as a long, slender projectile shot from a tube. But the early torpedo was basically a bomb on a stick. The term *torpedo* originally referred to any explosive device used in the water. The weapon got its name from the torpedo fish—a cousin of the electric eel—that attacks its prey by zapping it quickly with a fin carrying an electric charge. (Torpedo fish, in turn, get their name from the Latin *torpere,* which means to be sluggish or numb—which is what happens to the prey when a torpedo fish administers its shock.)

If the torpedo was anchored in the water, it was what we today call a naval mine. This was what David Farragut was talking about when he shouted, "Damn the torpedoes; full speed ahead!" at the Battle of Mobile Bay. At that particular moment, Farragut was trying to get a line of ships past the guns of Fort Morgan and into the bay. One of the lead ships, the U.S.S. *Tecumseh,* had just struck a mine, and sank in about a minute—taking ninety-two men down with her in the process. Farragut was telling the line to keep moving—torpedoes be damned.

If a torpedo was fitted to a spar—a pole used to support rigging, and commonly found on ships in the 1860s—it was, naturally, a "spar torpedo." The idea was to sneak up on an enemy ship, stick the spar torpedo into its side like a harpoon into a whale, and then row like crazy to get out of the way. Assuming you got away, you would then trigger the torpedo with a lanyard from a distance of 100 yards or so.

This was how during the Civil War, Lieutenant George Dixon—an army officer, ironically enough—became the first submarine commander to sink a ship. The Confederate sub *Hunley* sank the *Housatonic,* a Union sloop, four miles off Charleston Harbor in February 1864. The *Hunley* itself went down in the process, probably either by being damaged by the explosion or swamped by a wave.[2]

The Federals returned the favor later that year through the exploits of a young Navy lieutenant named William Cushing. Cushing, a rogue's rogue, had been dismissed from the Naval Academy after setting a new record in accumulated demerits. Pleading his way back into the service as an "acting master's mate," he managed to work his way up the ranks to receive a commission.

During the Civil War, Cushing became sort of an early version of a Navy SEAL, or special ops commando, or action hero—take your choice. In what became his most famous mission, he led a fourteen-man crew in *Picket Boat No. 1* to sink the 370-ton Confederate ironclad *Albemarle*. The C.S.S. *Albemarle* was anchored under guard in a protected cove along the coast of North Carolina. Cushing's picket boat was armed with a spar torpedo. He later described the operation in a report to headquarters:

> The light of a fire ashore showed me the ironclad, made fast to the wharf, and a pen of logs around her about thirty feet from her side.
>
> Passing her closely, we made a complete circle, so as to strike her fairly, and went into her bows on. By this time the enemy's fire was very severe, but a dose of Canister, at short range, served to moderate their zeal, and disturb their aim. Paymaster Swan of the *Otsego* was wounded near me, but how many more I know not. Three bullets struck my clothing, and the air seemed full of them.
>
> In a moment we had struck the logs, just abreast of the quarter-port, breasting them in some feet, and our bows resting on them. The torpedo boom was then lowered, and by a vigorous pull, I succeeded in diving the torpedo under the overhang, and exploded it at the same time that the *Albemarle*'s gun was fired. A shot seemed to go crashing through my boat, and a dense mass of water rushed in from the torpedo, filling the launch and completely disabling her.
>
> The enemy then continued his fire at fifteen-feet range, and demanded our surrender, which I twice refused, ordering the men to save themselves, and removing my own coat and shoes. Springing into the river, I swam with others into the middle of the stream, the rebels failing to hit us.
>
> The most of our party were captured, some drowned, and only one escaped besides myself, and he in a different direction. Acting

Masters Woodman of the *Commodore Hull*, I met in the water, half a mile below the town, and assisted him as best I could, but failed to get him ashore.

Completely exhausted, I managed to reach the shore, but was too weak to crawl out of the water until just daylight, when I managed to creep into the swamp, close to the fort. While hiding a few feet from the path, two of the *Albemarle*'s officers passed, and I judged from their conversation that the ship was destroyed.[3]

In other words, the *Hunley* was predecessor of the modern submarine, *Picket Boat No. 1* was the predecessor of the modern destroyer, and clearly, the most important component of the bomb-on-a-stick spar torpedo weapon system was machismo.

Robert Whitehead, an expatriate British engineer living in Austria-Hungary, came up with a major improvement in the torpedo—getting rid of the stick—a few years later. As the Americans were winding down the Civil War, Whitehead was busy designing steam engines for the Austrian navy. When Prussia took on Austria in the Schleswig-Holstein War of 1866—sort of a grudge match to decide who would be the alpha German in central Europe—Italy sided with the Prussians, with the stakes being control over the eastern Mediterranean.

When the Austrian and Italian fleets met off the island of Lissa in July, no one really knew what to expect. It was the first battle ever between two fleets of steam-powered, armored vessels. The Austrians were outgunned, but won the day by ramming their Italian opponents, running them down with propulsion provided courtesy of Whitehead's engines.

Whitehead became a hero in Austria, and his reputation soared in the rest of Europe too. With more money and political clout, Whitehead now had the opportunity to pursue his real love—not steam engines, but self-propelled torpedoes.

The idea of an "automobile torpedo" had occurred to others too. But Whitehead, unlike his competitors, had figured out how to make a torpedo that would run deep enough to hit its target below the waterline, but not so deep that it passed completely under the ship. He called his secret device, aptly enough, "the Secret."

The Secret was an adjustable hydrostatic valve that measured the water pressure outside the torpedo. A torpedoman would set the Secret to the desired depth and fire the torpedo. If the torpedo began to run too shallow or too deep, the Secret would detect the change in pressure and adjust fins at the weapon's tail accordingly.[4] This new breakthrough technology was so significant that before Whitehead would tell you how it worked, you first had to take an oath and sign a pledge not to reveal the secret about the Secret. In effect, Whitehead had created the first special access program, or, as it is called today in the defense community, SAP.*

The new torpedo was a huge commercial success, and Whitehead sold them by the thousands to navies around the world. Even so, the original Whitehead torpedo had a major shortcoming: the Secret could control its movement up and down, but not right and left. A ship would launch a torpedo several times in trials to discover its handling characteristics, and keep the results in record books. When you launched your torpedoes in battle, it was like sending familiar friends off to war. But it was hard to hit a target beyond a quarter mile.

It took another decade to fix this. In 1875 an American naval officer, James Adams Howell, replaced Whitehead's compressed-air motor with a 130-pound flywheel mounted fore-and-aft inside the weapon. The torpedoman would spin up the flywheel to 13,000 rpm—

---

* The Austro-Hungarian navy was one of Whitehead's first torpedo customers. It assigned a dashing young submarine commander, Captain Georg von Trapp, to Whitehead's factory near Trieste in 1908 to check on the production line. The next year he met Agathe Whitehead, Robert's granddaughter, during the christening of the U-5, a new submarine. Agathe and Georg fell madly in love and married in 1912. Robert Whitehead died in 1910, and von Trapp went on to a distinguished record in World War I—in the U-5.

When Agathe became ill with scarlet fever and died in 1922, von Trapp inherited her part of the torpedo fortune—a good thing, because after World War I the Austro-Hungarian empire was dissolved, and Austria no longer had a coast, or a navy, or submarines. Von Trapp remarried in 1927, this time to Maria Kutscher, a governess he had hired to help tend to one of his nine children, who was bedridden with rheumatic fever. In 1932 von Trapp was wiped out by the collapse of the Austrian National Bank, and later fled to escape the Nazis with Maria and the children.

Needing to earn a living, the von Trapps decided to go into show business. They eventually got to the United States. Maria recounted the whole story in The Trapp Family Singers, which became a perennial favorite among children's books. The book, in turn, became the story for the Rodgers and Hammerstein Broadway production The Sound of Music. In the movie version, Julie Andrews played Maria; Christopher Plummer played von Trapp.

it was like winding a clock—and the flywheel would power the torpedo for a range of a half-mile or so. More important, the flywheel kept the torpedo pointing dead ahead, much as a bicycle wheel keeps a bike upright and on course once it is moving.[5]

Soon after Howell introduced his torpedo, Whitehead came up with an even better solution. Instead of using a big flywheel to keep the entire torpedo properly aligned, Whitehead used a small flywheel mounted in a swiveling cage to detect when the torpedo strayed off course. That is, he used a gyroscope.

Gyroscopes had been around for almost a century, but they were mainly curiosities—scientific toys for demonstrating the laws of physics. Spin the wheel of a gyroscope on a tabletop, and inertia keeps it in a constant plane, even if the table is tilted, which is why guidance systems that use this principle are called inertial measurement units.

Whitehead connected his gyro to a rudder at the tail of the torpedo with a mechanism that was a masterpiece of Victorian design. The gyroscope connected to a lever, which connected to a valve that controlled a gas line from a tank of compressed oxygen. So whenever the torpedo began to stray off course, the gyroscope swiveled left or right, moving the lever, which activated the valve, which allowed the compressed gas to fill a small piston-and-cylinder device that operated a crank controlling the torpedo's rudder.

Whitehead needed this Rube Goldberg arrangement because a gyroscope produces only a tiny amount of force as it rights itself. It takes several pounds of torque to move the rudder of a torpedo moving through the water at twenty or thirty knots. The compressed air powering the rudder mechanism was able to overcome the force of water as the torpedo raced to its target. An engineer would say the linkage produced a "mechanical advantage." But this step was significant for an even more important reason.

The movement of the linkage between the gyroscope and the rudder contained *information*. The gyroscope was "telling" the rudder how many degrees to move to correct the course of the torpedo. Unintentionally, Whitehead had designed the equivalent of a simple, highly specialized mechanical analog computer. The linkage was doing multiplication. If, say, one degree of gyro movement produced two degrees of rudder movement, the system was multiplying a unit of gyro movement by two.

At the time, all calculators were mechanical, and this was how they worked—adding, subtracting, and multiplying by counting digits as the rotation of a gear or the ratcheting of a sprocket. They looked like small machines; Charles Babbage, who designed the granddaddy of all computers in 1837, even called his creation an "analytical engine." From some angles it looked like a five-speed transmission.

The point is that, for the first time, there was information moving around inside a weapon through a feedback loop. The gyros detected deviations from course and fed a mechanical signal to the control vanes. As the gyros detected that the torpedo was correcting its course, they returned the vanes to their neutral position.

The ability to use information in a feedback loop made the torpedo deadlier. In fact, adding a little information was much more effective than adding a lot of explosive. On the other hand, now the torpedo was totally dependent on information. If an adversary could just figure out how to get inside that feedback loop, he might be able to neutralize the weapon completely—Tom Rona's observation a century later.

Meanwhile, in New York, another inventor was working on a different problem with a similar solution. Shipbuilders were replacing wood with steel, and adding electric lighting to their vessels. All of these alien forces could perturb a magnetic compass. Elmer Sperry proposed using a gyroscope to ascertain changes in a ship's direction, thus creating a "gyrocompass."

In 1905 Sperry used his gyrocompass to design a robotic helmsman—it came to be called Metal Mike—that would keep a ship on a steady heading. This caught the attention of the U.S. Navy, which was interested in automating the control of ships. As it happened, the Navy was also interested in automating the control of the newfangled device the Wright brothers had demonstrated just two years earlier, the airplane.

Alas, flying is a three-dimensional problem and Metal Mike, for all his virtues, was just a two-dimensional character. So Sperry took the basic idea one step further and combined three gyros in a single unit he called an autopilot. The autopilot was linked to the airplane's ailerons, rudder, and elevator via levers and cables, and could keep an aircraft straight and level. This was handy if a pilot had to concentrate momentarily on something other than flying.

It did not take long for people to realize that if you could do without the pilot for a moment, you might just be able to eliminate him completely. In 1915 the Navy gave Sperry Gyroscope Company a $3,000 contract to build an "Aerial Torpedo," a catapult-launched N-9 Navy biplane with an autopilot filling the cockpit. The autopilot would keep the Aerial Torpedo on course, straight and level, and a timer would shut off the engine at the proper moment to drop the missile on its target.

The whole contraption was ready for testing off Long Island in September 1918, just as World War I was drawing to a close. The test crew aimed the weapon at a target eight miles away. The catapult hurled it into the air. The autopilot took over and put it on course. And then the timer failed. The Aerial Torpedo disappeared over the horizon, flying—straight and level—at an altitude of 4,000 feet.[6]

Autopilots became commonplace by the 1930s. But a few problems remained. First, mechanical autopilots were expensive, handcrafted instruments, something like a Swiss watch, not exactly what you would want to use in a missile designed for a one-way mission if you could help it. A guidance system has to be cheap if one hopes to have enough weapons for a war, and simple if one expects to build them in time. This was to become a major factor in the adoption of all guided weapons.

Second, the autopilot solved the problem of stabilizing a vehicle. This improved your chances of hitting a fixed target. But if the target moved, you were out of luck. Guided weapons needed some means to update their feedback loop if their targets tried to evade them, and this update had to arrive faster than the time it took for the target to take evasive action.

This was where a young glider pilot came up with the idea of using electronics. One of the fun tricks of flying gliders is called "riding the wave." You steer into a sixty-m.p.h. headwind coming up and over a nearby mountain range and adjust your indicated airspeed to sixty miles per hour by trimming the glider's "angle of attack"—that is, pulling back on the stick to raise or lower the nose of the aircraft. When you match your airspeed to the speed of the headwind, you will remain over one spot on the ground, but, magically, your glider will lift straight up a couple of thousand feet in just a few seconds, as though you put the aircraft on an elevator.

This is because, as any pilot knows, the airspeed indicator does exactly that—indicates airspeed. It doesn't care if you are making forward progress or not. This annoyed Helmut Hoelzer, an engineering student at the Technical University of Darmstadt. Hoelzer was learning how to fly gliders under a program promoted by the German government to get around the Versailles Treaty, which prohibited Germany from training pilots for an air force.[7]

Hoelzer designed a device that would measure his *absolute* velocity. It consisted of a spring-loaded ballast in a tube. A mechanical arm attached the ballast to a condenser that controlled an electric current passing through a circuit. If you put this apparatus in a vehicle and accelerated, the ballast would appear to move in the opposite direction (actually, the ballast remained stationary; it's the rest of the apparatus and the aircraft that moved). The arm would readjust the condenser, which would allow more electric current through the circuit. Measure the current, do some calculus, and you know your absolute velocity.

Today we would call this device an electronic accelerometer. Hoelzer thought it would be a great school project, only to have his professors dissuade him, saying it was too ambitious. But he did not give up. After he graduated, Hoelzer was drafted to take part in Wernher von Braun's missile project. Von Braun called his missile the A-4; Adolf Hitler later renamed it the *Vergeltungswaffen-2*—the V-2.

Hoelzer proposed using electronics to control every aspect of the missile's flight. Part of the system was essentially a version of the beam system the Brits learned to confound German bombers during the Battle of Britain. Instead of a pilot listening for a signal in his headphone, the missile used the signal to measure how much the missile stayed off course, and whether it was maintaining a proper attitude (not psychologically; whether it was pointing in the right direction).

This created a new problem, though. The missile could deviate from the beam in three directions (up/down, left/right, and forward/back), and deviate from its correct attitude in three axes (pitch, roll, and yaw). The rocket had four control vanes in its motor; each could be turned like a rudder on a ship to adjust the direction of the exhaust plume. So six different streams of information had to be combined to calculate the correct deflection for the exhaust plume, and then split into a separate signal to be sent to each of the four control vanes.

Hoelzer called the resulting electrical apparatus the *Mischgeraet*— a "mixing computer." It was, as far as anyone can tell, the world's first electronic analog computer. The cost savings were significant. Precision gyros such as those used in a mechanical autopilot cost $7,000 dollars. The simple electrical components that the *Mischgeraet* required to replace them cost just $2.50.

After World War II, Hoelzer came to the United States with von Braun. He worked for the Army for a while and, later, NASA. The designs he brought with him became the basis of the computers used in most early Army missiles—and the Saturn rockets that carried American astronauts to the moon.

Now there wasn't just information moving through a weapon. This was *electronic* information, traveling at the speed of light in an automated cycle similar to an OODA loop. The V-2's brain observed where the missile was and where it was pointing, compared this to where it was supposed to be, calculated the necessary correction, and sent the commands to the vanes that would make the necessary course correction. It was the same concept as Whitehead's Secret and Sperry's Aerial Torpedo, with electrons replacing the levers and control rods.

Even today, most guided weapons use the basic principles first demonstrated in the V-2. They all measure some signal that is related to the location of the target. It can be a radio beam indicating a path to the target. Or it can be a radar or laser beam that is reflected off the target. Or it can be an optical or infrared image of the target, or a satellite beacon. And—to complete this evolutionary tale—today even some artillery shells contain guidance systems. In each case, a signal is used as part of an electronic feedback loop so that the weapon adjusts its course until it is on target.

Once you had an electronic guidance system that took a weapon to a target, it was just an incremental step to follow a moving target. You simply adjusted the beam, tracking the target like a spotlight. The missile would either ride the beam, or home in on the beam reflected off the target.

So, by 1972, the United States was ready to use the first laser-guided bombs in Vietnam. One of the first targets was the Thanh Hoa Bridge, the same bridge where seven years earlier the F-105 flight leader had saved himself by reaching back into his brain for the move Captain John Boyd had told him about. The bridge had already survived 871 sorties and had cost the Air Force and Navy eleven aircraft.[8]

But this time was different. An Air Force fighter released a single laser-guided Paveway bomb. A spotter in another aircraft "painted" the target with a handheld laser. The bomb glided nicely to a spot illuminated by the laser on the center span, and the bridge was put out of service for months.

Chapter 10

A MATTER OF LATITUDE

Norman Schwarzkopf recalls Colin Powell telling him during Desert Storm, "I hope you realize that every time you fire another one of these cruise missiles, you know, it's two million bucks flying off into the air."[1] Powell was referring to the Navy's Tomahawk missile, a direct descendant of the Aerial Torpedo the Navy developed in 1918. Like the Aerial Torpedo, the Tomahawk is an unmanned aircraft fired from sea to attack a target on land. It had been key to U.S. success in the Gulf War, and was a versatile weapon, too, capable of being fired off either surface ships or submarines.

But Tomahawks use an especially complex guidance system—in fact, two systems. A Tomahawk is launched so that it flies over several pre-surveyed tracts of land en route. A radar in the missile compares the contours of the land with a terrain map in the missile's digital memory. A computer corrects the course of the missile. Then, as the missile approaches its target, an optical sensor looks for landmarks nearby and the computer makes a final course correction.

The Tomahawk's sensors are different from a beam-riding missile or laser-guided bomb, but the principle is the same: feedback, calculation, correction. The result is an incredibly accurate weapon that can attack targets from very long ranges, up to 1,500 miles. But, as Powell was noting, they are not cheap. That was the problem with precision weapons in the early 1990s. They cost too much. A laser-guided bomb cost much less, thanks to its simpler guidance system and not needing a motor—about $55,000. But this was still pricey.

Throughout the 1970s and 1980s, defense pundits argued about whether guided weapons were really as effective as claimed. The Gulf War seemed to settle the issue when precision-guided weapons worked so well that smart bombs became an icon of the conflict. (One joke at the time: "What's the difference between the typical smart bomb and the typical high school student?" Answer: "The smart bomb can find Baghdad on a map.")

After that, the main problem was simply building enough of the weapons to conduct a full-scale war. Despite all the publicity they received, guided munitions were just a small part of all the bombs the United States used in Desert Storm. According to Lieutenant General Buster Glosson, who supervised the planning of the air campaign, of the 85,000 tons of bombs used in the Gulf War, only 8,000 tons—less than 10 percent—were smart bombs. Even so, they accounted for nearly 75 percent of the damage.[2]

The main problem was price and the fact that they required so much labor to build. The solution was cheaper computer chips and consumer electronics. The price of microprocessors fell throughout the 1970s, 1980s, and 1990s, a result of Moore's Law, the totem of the Information Age economy. Exactly as Gordon Moore—one of the founders of Intel, the microchip giant—had predicted, the number of transistors a microprocessor maker could put on a chip grew exponentially over time, increasing computing power and decreasing its cost at a similar rate. The price of all electronics plummeted. At the same time, as consumers began to use more electronic gizmos at home, the military services were able to piggyback, spreading the costs around a much wider population and dropping prices even more.

So, when General Merrill McPeak jotted a memo to his executive officer on May 1, 1991, "We need to lay down a requirement for an all-wx PGM," the result was a much, much cheaper smart bomb.[3] McPeak was Chief of Staff of the U.S. Air Force. His memo was Air Force–speak for "I want a smart bomb that works in any weather." McPeak had stepped into his position just as Desert Storm began. His predecessor, General Mike Dugan, had made the mistake of talking too freely with the press, and Secretary of Defense Dick Cheney fired him.

The war had preoccupied McPeak for months, but now he had a chance to turn to new projects. Besides being expensive, laser-guided weapons had another drawback: You had to steer the bomb by keeping the laser trained on the target until impact. If smoke or fog obscured the target or blocked the laser during the minute or so the bomb needed to reach its target, the smart bomb would regress into stupidity. For that matter, if a weapons officer got confused and happened to focus the laser on the wrong object, the not-so-smart bomb would obediently follow.

But another factor was a revolution in a particular segment of information technology: navigation, or the ability to know where you were at a particular instant in time. The problem was mainly longitude or, more precisely, the inability to measure it. This may seem trivial today, but longitude is one half of any set of geographic coordinates. Without a definition of longitude and a way to measure it, modern navigation would simply not exist.

People understood what longitude was long before they knew how to determine their own longitude at a given instant. Unlike latitude (which is easy to calculate from the elevation of the North Star in the sky), longitude presents a real challenge, both to define and to measure.

To define the longitude of any location, you must first know the circumference of the earth. The mathematician Eratosthenes of Cyrene accomplished this sometime between 240 and 194 B.C., using a remarkable approach combining observation, logic, and a few simple measurements.[4] Eratosthenes was the keeper of the ancient library at Alexandria, in its day the most complete collection of records in the world. Back then science and religion overlapped; the stars and the earth provided both practical knowledge and predictions regarding human affairs. Geography was useful for locating distant lands, navigation, and gaining insight into science, music, and the gods—all of which were part of the same piece.

To estimate the world's circumference, Eratosthenes started by measuring the length of shadows cast in the midday sun in Alexandria and Syene (present-day Aswan). He understood that the rays of the sun are, for all practical purposes, parallel (the sun is very, very large relative to the earth, and very, very far away). Using the length of a stadium as a measure, he assumed the distance between the two cities equaled 5,000 stadia, based on the time a camel caravan required to travel from one city to the other. He also accepted the assumption commonly held at the time that the cites were on the same longitude.

Neither of these last two assumptions is entirely true, but they are close enough. Put it all together, and Eratosthenes knew two angles and the base of a right triangle, where the base runs along the surface of the earth and the point of the triangle is at the earth's center. With these three data and elementary geometry, you can calculate the radius of the earth and, from that, the circumference.

Eratosthenes estimated the total distance around the earth was 250,000 stadia. This equaled about 25,000 miles, which was remarkably close to the earth's true circumference of 24,899 miles. It was a rough estimate, and even today experts argue about its accuracy. For one thing, there was no standard length of a stadium; the arenas at Rome, Athens, and Alexandria, for instance, are all slightly different in size. But Eratosthenes was in the ballpark (so to speak).

Once you know the circumference of the earth, it is easy to divide this number into equal units. Turn the measure ninety degrees so it runs parallel to the equator, and you have longitude. At least it was a start, and Eratosthenes gets credit for the first map using a close-to-accurate grid of longitude and latitude.°

Still, even if learned people understood what longitude was, centuries went by before anyone figured out a practical way to measure it. So it was impossible to navigate freely over the earth's surface. On land, explorers stuck to trails. At sea, sailors clung to coastlines or tried to measure their progress east or west by the number of days they had sailed since losing sight of land. Navigation was hit-or-miss. Or, as Benjamin Franklin recalled from his own voyage from New York to London in 1757:

> We were several times chas'd in our passage, but outsail'd every thing, and in thirty days had soundings. We had a good observation, and the captain judg'd himself so near our port, Falmouth, that, if

° In addition to discovering longitude, Eratosthenes also invented the prime number sieve, a process for compiling a list of numbers that cannot be produced by any set of multiples other than itself and 1. Prime numbers are used today to formulate cipher keys. So, just as Albert Myer can be considered the distant father of NSA and NOAA, Eratosthenes can be considered the distant father of the two basic methodologies those two organizations rely on: factoring and geolocation.

we made a good run in the night, we might be off the mouth of that harbor in the morning, and by running in the night might escape the notice of the enemy's privateers, who often crus'd near the entrance of the channel. Accordingly, all the sail was set that we could possibly make, and the wind being very fresh and fair, we went right before it, and made great way. The captain, after his observation, shap'd his course, as he thought, so as to pass wide of the Scilly Isles; but it seems there is sometimes a strong indraught setting up St. George's Channel, which deceives seamen and caused the loss of Sir Cloudesley Shovel's squadron.[5]

Franklin did not have to identify Admiral Shovel because all of his readers knew exactly who he was. Fifty years earlier, on October 22, 1707, Shovel was leading a squadron of about twenty ships back to Britain after laying siege to Toulon. The squadron lost its bearings, thought it had a clear shot into the English Channel, but ran aground on the rocks off the Isles of Scilly, a patch of islands about twenty miles off Land's End. Shovel and the ships' masters had miscalculated their position by almost fifty miles.

Four ships and around 2,000 men (including Shovel) were lost. It was a calamity on the order of the explosion of the space shuttle *Challenger*, combined with the sinking of the *Titanic* and the death of a top military commander with the standing of, say, a Schwarzkopf or Powell mixed in. Bad navigation had caused thousands of deaths at sea over the centuries, and kings and their governments had offered prizes to whoever could find a way to measure longitude, starting with Philip II of Spain in 1567.

The challenge in calculating longitude is that although the heavenly bodies clearly appear to move as you travel east or west, the exact pattern is a function of several factors. The earth's rotation changes the position of the stars in the sky during the night. The earth's orbit changes the positions of the sun and stars over the course of the year. And the moon has its own orbit.

The challenge is more a matter of practical engineering than theoretical science and consists of three problems. First, you need to know your exact time. Second, you need to know the location of the sun or moon as you see it in the sky. And third, you need to match that observation to the one you would expect to see if you were in a given location at a given time; that's your longitude.

The British began working on the last part of the problem, compiling astronomical tables, in 1675, when Charles II founded the Royal Greenwich Observatory. Charles instructed John Flamsteed, the Astronomer Royal, "to apply himself with the most exact care and diligence to the rectifying of the tables of the motions of the heavens, and the places of the fixed stars, so as to find out the so much desired longitude of places for perfecting the art of navigation."[6]

Seven years after the Shovel disaster, Queen Anne signed the Longitude Act, offering a prize of 20,000 pounds as a "publick reward for such person or persons as shall discover the Longitude." This act established the Board of Longitude, which was the first science-funding body in Britain.[7] Think of the Royal Greenwich Observatory as the distant ancestor of today's National Institute of Standards and Technology. The Board of Longitude could be thought of as the ancestor of the National Science Foundation.

In 1731 John Hadley, a mathematician and a member of the Royal Society, solved the problem of measuring the location of objects in the sky with a device he called an octant. Hadley's octant was a neat, compact apparatus consisting of mirrors, a pivoting sighting scope, and a protractor-like scale that measured the elevation of the scope from the horizontal. The scale on Hadley's octant had a range of 90 degrees; when the design was modified to a more useful 120 degrees, the result was the familiar sextant.

A sextant works by triangulation, the principle that says if you know the location of three points and your distance from each, you can calculate your own location using simple geometry. The optics in a sextant let you sight the horizon and the sun, moon, or a star simultaneously. The scale indicates the apparent altitude of the object in the sky. Since we know where these objects are supposed to be at any given location at any given time on any given day, a sextant sighting, combined with a good timepiece and a nautical almanac, tells you where you are with reasonable accuracy.

That left the problem of measuring time. The first hurdle was to find a natural phenomenon that occurred in a regular, dependable cycle—in effect, a mechanical representation of a fixed unit of time.

Galileo had already thought of using a weighted pendulum, which naturally swings back and forth at a regular rate, but never got around to building a working model. Being imprisoned by the church for arguing the heavenly bodies moved (which was essential to the whole idea of calculating longitude anyway) got in the way of his work.

So it was left to Christiaan Huygens to build the first pendulum clock in Holland in 1656. The problem then was figuring out how to make one that was rugged enough to put on a ship. The clock also had to keep reliable time, no matter the weather—rolling waves, freezing temperatures, blistering heat, or anything else.

John Harrison, a joiner from Lincolnshire, was a self-taught clockmaker who solved the portability and reliability problems. In 1713 Harrison built his first clock, essentially what we would call a grandfather clock. After that, Harrison had to figure out how to make the pendulum mechanism smaller and independent of gravity. During the next sixty years, Harrison slaved at the problem until finally, at the age of seventy-nine, he won the prize with a device that resembles a modern pocket watch.[8]

If it were not for Eratosthenes, Hadley, and Harrison, not only would there be no modern navigation, there would not be any precision-guided weapons. Being able to define and locate a set of coordinates is what makes zapping possible.

Instead of laser guidance, the new weapon that was developed to answer McPeak's memo uses the Global Positioning System (GPS), a satellite navigation technology. A GPS receiver can tell you your location within a meter or so by interpreting signals from a constellation of satellites. The system works by triangulation—there's that principle again—calculating your location from the known location of three points and simple geometry.

GPS uses this principle twice. First, each satellite calculates its own location from radio signals transmitted from earth stations whose locations are precisely known. Then the satellite broadcasts its own location to earth, along with an additional piece of key information: a time stamp from an on-board atomic clock, which, like all such devices, can measure time with incredible precision.

We know that the signal from the satellite travels at the speed of light, and we know from Albert Einstein that this speed is a constant value. Since the time stamp is accurate to a billionth of a second, the signal can tell you your distance from the satellite. So when you receive a signal from three GPS satellites, you have three lines of precisely known length to precisely known points in space, which is what you need to calculate your location.

Like smart bombs, GPS became a technology star in Desert Storm, thanks to its role in the Left Hook. Just as ships had been getting lost at sea for centuries until modern navigation came along, caravans had been getting lost in the desert. With GPS, the Coalition forces could race into the desert, and even in the dark and in sandstorms, they knew exactly where they were instantly. (The Pave Low helicopter that guided Task Force Normandy was equipped with GPS.)

The only problem was getting enough military-specification GPS receivers for all of the troops. But that is where the commercial connection comes in, and one reason why precision-guided munitions are so cheap today.

Once GPS satellites were in orbit, companies began to make GPS receivers for civilians. Backpackers and hunters used them to find their way in the great outdoors. Truckers and railroads used them to track freight. By 1990 you could buy a handheld GPS receiver from a company like Trimble or Magellan for just a few hundred bucks, about what you would expect to pay for a device that is basically just a radio receiver and a calculator.

Soldiers started writing home telling their folks how wonderful GPS was—and complaining that there were not enough receivers to go around. Parents began driving down to the local Radio Shack to buy commercial GPS receivers, and would mail them to their kids in Saudi Arabia. Word got around the troops: "Uncle Sam can send you off to war, but you still need Mom to help you find your way home."

McPeak merely wanted a bomb that worked in the rain—GPS does not care whether you can see your target—but there was an important bonus. By the mid-1990s, electronics companies were making GPS receivers by the tens of thousands. Thanks to economies of scale, they became really inexpensive.

The weapon that eventually answered McPeak's memo—the Joint

Direct Attack Munition (JDAM, or "jay-dam")—consisted of just a GPS receiver, a miniature inertial measurement unit, and a set of control fins. You could bolt the JDAM to the back of a World War II or Korean War dumb bomb, and have a smart bomb that cost just $16,000.

That was critical to making precision strike practical: getting the cost down, and getting production up. Even before the JDAM program was underway, some of the Pentagon's thinkers could see where the trends would lead. This included, of course, Andy Marshall.

Marshall had been following precision strike since the late 1970s, when the Army began developing a never-to-be-deployed weapon called Assault Breaker to defend against a Soviet invasion of Western Europe. U.S. analysts believed they had a good understanding of the Soviet invasion plan to invade the West if only because the Soviets practiced it every year. The Soviet plan was a traditional attack, similar to what the Germans had used in World War I and World War II, except starting from a point farther east. The Soviets planned to launch a first wave with the forces they kept in East Germany, Poland, and Czechoslovakia. These forces would probably take a lot of losses in trying to break through NATO's defenses. So the Soviets planned to follow up with a second wave about two weeks later, with forces based in the Soviet Union. This would be followed by a third wave several weeks after that.

The later waves were to consist of reserve units. The Soviets planned to call former draftees back to duty, get their equipment out of storage, and send them to the front. So a NATO defender who managed to hold his ground at, say, Fulda, would first see a wave of twenty-year-olds, then a wave of thirty-year-olds, and then a wave of forty-year-olds, driving equipment that was ten, twenty, and thirty years old, respectively.

This was a typical war plan for a mechanized army in the twentieth century, and its prospects for success reduced down to simple mathematics. To win, according to the experts, the Soviets would need to throw enough men and machines on the Central Front so that they had a 1.5 to 1 advantage over NATO. If they could do that, then theoretically they could mass their forces and break through NATO lines.[9]

Assault Breaker was supposed to destroy these reinforcements before they reached the front. According to the plan, NATO forces would fire Assault Breaker missiles fifty to a hundred miles behind the front line over the Soviet reserve forces as they were preparing for attack. Each missile would release dozens of heat-seeking bomblets that would home in on the engines of the Soviet tanks. Since the reserves would never reach the front line, the Soviets would never achieve the magic 1.5 number, and the attack would fail.

The Army abandoned Assault Breaker, but the exercise got Marshall thinking about how this kind of new technology would affect the future of warfare. As it happened, the Soviets were writing about what Marshal Nikolai Ogarkov had called a "military-technical revolution." Ogarkov, who was Chief of the Soviet General Staff, warned his fellow officers that high-tech weapons were about to upset the entire balance of power. Marshall appropriated Ogarkov's concept, and wrote a memo for the incoming Clinton administration describing the impending "Revolution in Military Affairs." The lynchpin that held the idea together was information technology—in particular, precision strike.[10]

By the time the United States fought in Kosovo in November 1999, the Defense Department had enough of the smart bombs to carry out an entire campaign based on precision strike. Smart bombs made up almost two-thirds of the weapons used—22,000 weapons—and were hitting their targets about half the time. The hit rate two years later during Operation Enduring Freedom in Afghanistan was even better, about 75 percent.[11]

Programming a JDAM was like addressing a FedEx envelope; you just type in the GPS coordinates, and the bomb goes where it is told. Assuming you knew the location of a target, you could send the weapon to it, and be certain that it would arrive. A single B-2 stealth bomber could carry sixteen JADMs, program each to seek a different target, and release them in a volley so that the entire set of targets would be suddenly—and nearly simultaneously—destroyed. So it might seem as though we have achieved the Holy Grail—low-cost, long-range weapons that hit their target most of the time. But, like most matters having to do with war, the situation is more complicated.

During the Gulf War, commentators liked to remark about how combat seen through night scopes and fought with smart bombs seemed like a video game. Anyone who was on the scene would likely disagree, unless your video arcade happens to be in a free fire zone. And even smart bombs can kill civilians when civilians are mistaken for combatants or when soldiers use them as human shields. And the struggle between offense and defense continues even in the Information Age. Our opponents will work on countermeasures—GPS jammers, devices that dazzle laser seekers, and the like. If we depend heavily on smart bombs and our opponents find the right countermeasure, the result could be really, really bad.

Besides, precision strike is just one option in network warfare. The other is swarming, and one can be as effective as the other—and, thanks to modern information technology, an adversary that lacks the money to buy precision weapons can use this second option to compensate. This became clear in Yemen, nearly a year before the attacks of September 11.

Chapter 11

SWARMING

On the morning of October 12, 2000, the U.S.S. *Cole* arrived at the naval base at Aden, on the southern coast of Yemen. Until 1967 the base on the Arabian Sea had been a British outpost, built to protect the trade routes to India and the southern approach to the Suez Canal. For many years Yemen was probably best known as one of the world's leading producers, and consumers, of *khat,* a mildly psychotropic leaf that gives you a buzz if you chew a wad of it. Yemen has barely qualified as a state in the modern sense of the word. Tribal chiefs and religious leaders in the countryside have usually exercised as much control as the government, and even when the British occupied Aden, their influence did not go far beyond the city limits. For that matter, until 2000 Yemen's borders were not even formally defined; it was simply understood that there was just some point in the vast desert of the Empty Quarter where Yemen ended and Saudi Arabia began.

The *Cole* was en route to the Persian Gulf and anchored in Aden harbor for a six-hour gas-and-go stop. The Pentagon had signed an agreement with Yemen to use the base as a refueling station. The Clinton administration hoped an infusion of cash from the United States might improve relations with the Yemeni government. This, in turn, would help U.S. efforts to build a coalition to oppose Saddam Hussein in Iraq and Islamic fundamentalists in Iran.

Many boats were moving about the harbor that morning. Just before noon, one of them, an inflatable skiff, turned toward the *Cole* and began to approach the destroyer. Some of the American sailors thought the skiff was bringing supplies. The two men on the skiff rose and waved as the boat continued straight toward the *Cole*. It looked like the boat was going to bump into the American ship amidships—a minor mishap. Just what you might expect in a Third World port.

But instead of merely bumping into the *Cole,* the skiff exploded in a tremendous fireball as it neared the side of the destroyer. The blast could be heard across the harbor and put a forty-foot hole in the ship at the waterline. The *Cole* immediately started to flood, and then began to list. The destroyer was close to sinking until the crew finally got the damage under control. Seventeen sailors were killed and another 39 wounded.[1]

It was another attack by Al Qaeda, and its method was essentially the same one that George Dixon used to sink the *Housatonic* and William Cushing used to take out the *Albemarle:* small boat, big explosive, determined men. In other words, after 150 years of effort to perfect guided weapons—getting rid of the damned stick—Al Qaeda showed you could still use an old-fashioned torpedo boat to sink an enemy warship. Al Qaeda's success still depended on information technology. But the technology was not in the weapon. Al Qaeda succeeded in assembling its forces, approaching the *Cole,* and striking because its *organization* had the information edge.

The attack on the *Cole* was a textbook case of swarming, the other mode of attack in network warfare. Our ship ventured into their network, unaware of the danger. They waited for the optimal moment, massed for the attack, struck, and then disappeared.

Arquilla and Ronfeldt began to think about swarming as something that went hand-in-hand with network warfare. The link between the two was information technology. The same communications systems that enabled virtual organizations to operate without a hierarchy or a geographical base also enabled them to mass an all-on-one attack from the ether. Thanks to the technology, everyone in a network could share information to develop a "common picture of the battle space" and coordinate actions without a single hand controlling every move.

Arquilla and Ronfeldt saw many analogies in nature. Bees swarmed. Wolves swarmed. Sharks swarmed. So did paparazzi. Nature had evolved its own "concept of operations," where animals worked in a loose organization with no permanently defined leader, exchanging cues and cooperating to hunt their prey. When the zebra was least expecting it, the members of the network would spring out of the grass, strike, and, when finished, disappear just as quickly.

In a swarming attack, members of the swarm are interconnected by communications. Each member occasionally "pings" its brethren to find out what they are doing and what information they have.[2] The members can be special operations forces, or they can be conventional ships, aircraft, and land vehicles. In any case, they keep a low profile to avoid detection. When they spot their target, they pounce to attack, possibly simultaneously from several directions or when the target is most vulnerable.

Arquilla and Ronfeldt started thinking about swarming as part of an Information Age organization. But some military thinkers had been working on swarming even before that. The idea seems to be an offshoot of thinking about "small wars."

Military writers have long tried to draw a line between major wars and small wars. In a major war, one army meets another in battle, usually to fight over a particular piece of territory. Major wars usually have clear battle lines separating the combatants. And there is usually a clear beginning and end.

Small wars, on the other hand, are messier, lesser conflicts. Often they are linked to ambiguous foreign entanglements with indefinite endpoints—peacekeeping operations, counterinsurgency, relief missions, and the like.

Military thinkers have drawn the line between major wars and small wars partly because each needs different capabilities. Major wars require big armies and navies; small wars require a complex mix of small combat forces, large military units, and police. In a small war, even diplomats can get into the thick of action because the lines between combat and negotiation are blurred.

It is little wonder military officials and professional soldiers have traditionally loathed small wars. In a small war, the adversary—or, even worse, adversaries—skirmish, snipe, ambush, and set off booby traps. Often there is no clear distinction between combatant and non-combatant. Irregulars—often anyone who can tote a gun or knows which side of a claymore mine is supposed to face the enemy—can join in. You are under constant risk because there is no front line. It is harder to identify the adversary trying to kill you, and easier to make a mistake and kill a noncombatant. It's tough to think of any small war that ended with a triumphal parade, partly because few small wars are popular, and partly because it is often hard to tell if the war is really over. Small wars can drag on indefinitely.

And there is the bottom line: For a soldier, a small war can be just as deadly as any major war.

Alas, military leaders might prefer to prepare for major wars, but a quick glance at history shows the United States has had to fight small wars at least as often. Consider the list: the wars with the Barbary States, the undeclared naval war with France, the various Indian wars of the 1800s, border skirmishes with Mexico up to the early 1900s, intervention in the Philippines, Haiti, and China in the early decades of the twentieth century, and so on.[3]

Such small wars were so frequent that in the 1930s the Marine Corps, which often had the job of fighting them, even issued the *Small Wars Manual*. The manual warned, "Small Wars demand the highest type of leadership. . . . Small Wars are conceived in uncertainty, are conducted often with precarious responsibility and doubtful authority, under indeterminate orders lacking specific instructions."[4] It was harder to plan a small war, so officers had to think on their feet. They had to improvise.

The Marines issued the final edition of the *Small Wars Manual* in 1940. Then small wars fell off the scope when the United States fought World War II and the Korean War. Small wars were largely neglected during most of the Cold War, as the United States devoted most of its efforts toward planning for the ultimate big war with the Soviets.

This was true until the collapse of the Soviet Union. Try as they might to focus on big wars, throughout the Cold War the U.S. military often had to act in small, off-the-cuff operations. Again, the Marines often had the lead, as when they landed in Beirut in 1958 and the Dominican Republic in 1964 to assist friendly governments that were threatened by domestic opponents.

The Marines developed a new interest in these new less-than-major conflicts. They were in the tradition of the *Small Wars Manual*. Besides, the Marines had a tradition of searching for and solving problems in specialized—call it "boutique"—warfare. It distinguished Marine missions from the large-scale, conventional operations that the other services engage in.

The Marines also had one very special interest in the problem. In the early 1980s Lebanon slipped into one of its periodic civil wars between Muslims, Christians, and Druze. And, for good measure this time, Palestinian fighters who had been booted out of Jordan back in 1970 also joined the fray. The United Nations had managed to negotiate an armistice, and the United States sent a contingent of Marines to help monitor the cease-fire.

On April 21, 1983, a truck wove through the guard posts protecting a building the Marines were using as a barracks outside Beirut. The truck plowed into the building's entrance and blew up, bringing down the entire concrete structure. It was the first major terrorist suicide bombing against a U.S. foreign base. No one officially claimed credit, but most American experts believe that the Iranian-backed Hezbollah organization was responsible.

The bombing killed 241 Marines, and more than 100 people were wounded. The Marines have fewer than 200,000 personnel total, so nearly everyone in the Corps probably knew one of the casualties, or at least knew someone who knew someone. The Marines began to study the problem of small wars—now called "low intensity conflict"—more closely.

As it happened, one of the people pushing the Marines in this direction was John Boyd. To understand how an Air Force fighter pilot came to shape Marine Corps combat doctrine, one first needs to backtrack through the seemingly unrelated topic of railroads.

In 1973 William Lind was looking for a job with Amtrak. Lind had graduated from Dartmouth a few years before as a history major. He had grown tired of graduate work and the liberal political atmosphere at Princeton, and decided to pursue his true love: trains.

Lind, a Cleveland native, wrote to his Senator, Robert Taft, Jr. (R-Ohio), asking for some help in landing the job. Taft couldn't help, but he was impressed enough by the young man's résumé to offer him a position on his own staff. Taft was a member of the Armed Services Committee. After he arrived, Lind asked the Senator if he had anyone covering defense; Taft said no, and Bill Lind became a defense policy analyst.

Taft was also a history buff, and he and Lind hit it off. They soon found themselves wondering why other Senators and staffers seemed only concerned with funding the next program based in their state, and uninterested in what seemed like the biggest issue facing the Armed Services Committee, which was how to prepare the United States to win wars.

Taft began the Defense Reform Caucus to look at these bigger issues, and that soon caught the attention of another Senator, Gary Hart (D-Col.)—at the time, a high-profile politician and a leading contender for the Democratic presidential nomination. Hart was looking for new ideas for U.S. defense policy.

It's hard to appreciate today, but the Democratic Party's defense policy in the late 1970s had slipped far, far to the left. There was a wing in the party that believed not only that the United States spent too much on defense, but that the U.S. military was *too* powerful, and that made war more likely. That's why they wanted to cut the defense budget by, say, half.

To get a taste of this thinking, consider this quotation from *Winding Down*, a text widely used in defense courses at leading universities in the early 1980s. It was considered liberal, but still within the range of reasonable opinion; the authors included Philip Morrison, an MIT professor and former Manhattan Project scientist who edited book reviews for *Scientific American*:

> Speculating about the future, we have some hope that a major U.S. cutback would not be merely a one-shot reduction, to be eroded by some inexorable global increase in armaments. Since World War II, the United States, more than any country, has led the world toward militarization. It has continually advanced the frontier of military technology. It has aided the expansion of foreign military forces and foreign defense industries. Most important, it has continually upgraded its own military forces to maintain the image and reality of military superiority over the U.S.S.R. (an edge which the Soviet

Union in turn consistently tries to narrow), rather than allow the U.S.S.R. to achieve something widely perceived as an even balance of forces.[5]

This may have appealed to the Brie and Chablis set who graduated from Bennington College back in '68 or '69 and got their insights into national security from Noam Chomsky in *The New York Review of Books*. But from the viewpoint of winning elections, it was a loser. Democrats like Hart knew they needed an alternative, a policy that they could sell to committed Democrats in the spring and summer primaries, but which they could also sell to everyone else in the general election in November.

That drew Hart to defense reform. He was naturally drawn to anything that seemed like "new ideas." In 1984 he ran promising exactly that. Defense reform was a good political posture, too. By supporting defense reform, you could oppose current policies, but you also had an alternative to offer in its place. It seemed like the sensible center. You were not *against* defense; you just wanted something *different*.

When Howard Metzenbaum defeated Taft in 1976, Hart kept Lind on. Lind began to publish articles on how to reshape the U.S. military. One of these, which argued that the era of the manned aircraft was over, found its way to a part-time analyst working in the Pentagon on the next-generation fighter—John Boyd.

Opinions on fighter aircraft aside, Boyd and Lind were made for each other: if Boyd could be considered unconventional, judgmental, or a firebrand, he had a soulmate in Lind. When you entered the staffer's office, a larger-than-life photo of Benito Mussolini greeted you from the wall. It had nothing to do with his job as a defense analyst; Lind, the eternal railroad buff, just admired the fact that Mussolini made the trains run on time.[*]

[*] Just to bring this full circle: After Gary Hart was pictured on the cover of the *National Enquirer* with aspiring model / former University of South Carolina cheerleader Donna Rice on his lap while aboard the yacht *Monkey Business,* his presidential aspirations flamed out.

· William Lind, deeply conservative on social issues, notwithstanding his work on defense policy with Hart, joined the Free Congress Foundation, which is headed by Paul Weyrich. Weyrich, it will be recalled, was the man singularly responsible for knocking off John Tower's nomination as Secretary of Defense, which, in turn, opened the way for Dick

During this time, Lind caught the attention of the Marines. Of all the services, the Marines are probably tuned in to Congress best. As Lind puts it, "The Marines' political base is the Hill." This is pure survival instincts at work. Being so small, the Marines are often pushed to make a case why they should not be folded into the Army. So the service keeps close tabs on the Hill, and if there is a staffer with a deep interest in nineteenth-century Prussian military tactics, the Marines are ready to listen to him, especially if he works for a high-profile Senator on the Armed Services Committee with an eye on the White House.

Soon Lind made the connection between the Marines and Boyd, bringing Boyd to the attention of Colonel Alfred Gray sometime in the late 1970s. Gray was a Marine's Marine—he had enlisted as a private and later advanced through the ranks to become a four-star general. But Gray also had an intellectual side and was sympathetic to gadfly historians like Lind, who were determined to make both Congress and the Defense Department question how America prepared for war. After Lind made the introduction, Boyd and Gray hit it off.

Soon the two men were debating which translation of Sun-tzu best captured the master's understanding of the art of war. Later, when Gray became Commandant of the Marine Corps, he instructed his staff to begin rethinking Marine doctrine to incorporate Boyd's thinking. The Marines issued a new manual in 1989 describing its approach to fighting wars titled, aptly enough, *Warfighting*.[6] The new manual was, in effect, the guide to war that Boyd never got around to writing.

Warfighting contained all of the ideas Boyd had pushed—in particular, "maneuver warfare." The basic objective in military maneuver, as we have seen, is to be agile enough and fast enough so that you can hit your enemy where it is weakest. The larger goal in maneuver warfare is to create disorder and continually keep the opponent off balance so that you defeat him quickly and efficiently.

If you begin to consider the possibility that your enemy might use

Cheney to become Defense Secretary and appoint Duane Andrews, who brought in Ron Knecht to push through the first policy directive on information warfare. Donna Rice, after making a television commercial for No Excuses Jeans, disappeared from public view until 1998, when she re-emerged as an advocate for protecting children from pornography on the Internet. Gary Hart has re-emerged as an expert on homeland security.

the same approach, you are left with two highly mobile forces, each trying to find weak spots in the other and make decisions faster than their opponent can respond. The Marines did not give up the idea of a hierarchical structure. But if you add a lot of networked communications, and assume a commander provides just his strategic intent to his troops and leaves most of the details to his forces, maneuver warfare begins to look a lot like swarming.[7]

And that is exactly the kind of combat operations we have been seeing lately. As we have already noted, Al Qaeda used network warfare to strike the *Cole* and in the September 11 attack. CIA paramilitary units and Army Special Forces, the first U.S. personnel to arrive in Afghanistan to help the Northern Alliance, used similar tactics. Small cells, dropped into the middle of hostile territory, coordinated their operations both with each other and with their commanders back home, thousands of miles away.

But the same ideas also work on a larger scale, and are quickly becoming the norm. Just over a month after the United States put its first forces into Afghanistan, it began to send in larger units, and stuck with the same approach, where units operated as part of a network and the idea of a "front" meant little in the traditional sense. It is even more telling that the first large units the United States sent were Marines. The new doctrine paid off. Forget the images of John Wayne leading waves of amphibious craft in beach landings in *The Sands of Iwo Jima.* Today the Marines are more likely to fight as a quickly deployed network of semi-autonomous units in a landlocked country hundreds of miles from the nearest ocean.[8]

The ideas and technology are irresistible. A year later, when the Pentagon began to prepare plans for military action in Iraq, there was a lot of disagreement over how big a force was needed. The uniformed commanders (typically) wanted as many troops as possible; civilian officials (as is also typical) thought the operation would succeed with a much smaller force.

The press made a lot of the disagreement at the time, but usually missed the more interesting fact. Despite the cliché about generals insisting on fighting the previous war, just about everyone knew the paradigm had changed. No one was proposing a replay of Desert Storm. The question was not over whether to win the information war and fight as an agile network deployed throughout the enemy's territory, but how best to do so.

Today combat may be on a small scale, like in Yemen. It may be in Pakistani towns, Filipino jungles, or desolate Afghan mountains. Or it may be a huge military operation involving hundreds of thousands of troops. In any of these scenarios, armies using the old tactics will be defeated, or will simply be left at home.

The hardest problem in swarming is how to exchange information and coordinate action. The solution came from—of all places—Japanese textile sweatshops.

Sweatshops have been the target of such social reformers and left-wing pundits as Charles Dickens, Friedrich Engels, Samuel Gompers, and most recently, Garry Trudeau. But the fact remains that just about every country that has an advanced manufacturing economy today has gone through a phase in which illiterate peasants left the countryside to work in sweatshops, making clothing for the masses.[9]

Japan was no exception, and got started in manufacturing via textiles in the early 1900s. It was following the same path Britain and the United States had traveled in the eighteenth and nineteenth centuries, and which South Korea, Singapore, Taiwan, and Malaysia would follow in the twentieth century.

In 1926 a businessman from Shizuoka prefecture, Sakichi Toyoda, took advantage of the exploding Japanese textile industry by designing a power loom that automatically shut down when a thread snapped. Toyoda was soon selling automatic looms around the world. However, Sakichi's son, Kiichiro, was more interested in cars than textiles, and convinced his father in 1936 to get into the automobile business. Kiichiro went to Dearborn to learn how to build cars from Ford. (It may seem amazing today, but in the 1920s Ford, and General Motors, were major exporters of cars to Japan.)

Ford had just completed its Rouge River works ten years earlier. "The Rouge" was a huge complex covering 2,000 acres where as many as 100,000 men might report to work in a day. Ford made virtually every part of a car at a single factory. Raw materials—iron ore, wood, rubber, coal—poured in one end of The Rouge, and Model A Fords rolled out the other. Kiichiro Toyoda was impressed. But he did not think it would fit in Japan, where space was tight and both politics and tradition protected small companies, like the ones that make starter motors, batteries, shock absorbers, and other automotive parts.

Kiichiro's solution was to keep his suppliers, but instead of stockpiling components at the Toyoda works, he made his suppliers agree to deliver their parts exactly when they were needed—just in time. This cut inventory costs. "Just In Time" became a basic principle in the science of manufacturing.[10]

During World War II American B-29s leveled most of the Toyoda factories, and after the war the Japanese government squeezed Kiichiro out of power by reorganizing the company. Eiji Toyoda, Sakichi's nephew, was named managing director of the new Toyoda Automotive Works, which he renamed Toyota (In Japanese "Toyoda" requires ten characters, while "Toyota" requires eight, which is luckier and more aristocratic).

When the Japanese economy took off in the 1970s and 1980s, Japanese management style—including Just In Time—became the vogue in America. Conservative businessmen liked it because Just In Time gave them a rationale to downsize, spin off, and outsource. Liberal pundits liked Japanese management style because it supported a larger role for government in economic planning. In other words, Gordon Gecko meets Michael Dukakis. As long as the Japanese economy was growing at twice the rate of the U.S. economy, nobody was going to argue.[11]

In the 1990s the Japanese real estate bubble burst and Japan's economy stalled out. No one wanted to hear about Japanese management style. But by then, Just In Time (which was, in its essence, just a sensible argument for optimizing centralization and decentralization) had taken hold—just in time for the Internet to take over America.

Just In Time and the Internet developed totally independently, but were a natural fit. Just In Time was the concept; the Internet was the technology that made it possible. The World Wide Web let you control and move inventories with a point and click. You could find out what your customer needed without a face-to-face conversation, and he could order from you without making an appointment.

Management thinkers in American business schools picked up

where the Japanese had left off. Just In Time plus Internet equaled e-commerce and New Economy. This new business concept also happened to address an urgent need of California's Silicon Valley technology developers: finding some way to knock off their arch-competitor, Microsoft.

By the early 1990s Bill Gates, former hacker/current billionaire, had pretty much crushed every competitor in the software business. Microsoft owned about nine-tenths of the market for personal computer operating systems, and was well on its way to capturing the market for applications like word processing, spreadsheets, and graphics.

Microsoft's competitors knew they were in a game they could not win under its current rules. Their only option was to try to start a new game, with new rules that put everyone on an equal footing. The new game was called network-centric computing. Its champion was Scott McNealy, chairman and CEO of Sun Microsystems. Sun manufactured servers, the computers that connect users to the Internet. Servers control incoming and outgoing traffic, and store what users see as a Web page in the form of files copied on the server's hard drive.

But for Sun and everyone else who feared Bill Gates, the most important thing about servers was that most of them did not use Microsoft software. If a new approach to computing depended on servers rather than PCs, there might be a chance to start a new competition in which Microsoft's huge lead in personal computer software was no longer important.

McNealy was ready to tell anyone who would listen that "platform-centric" computing—meaning, personal computers using Microsoft software—was passé. The Internet was the thing, he said. Computing would become *network-centric*. Developers would write programs for the Net, businesses would store their records on the Net, and people would do their computing on the Net. The desktop PC, McNealy said, would eventually atrophy to a ghost of its present self.

McNealy might have been viewing the world through a Sun-centric lens, but, in fact, the Internet was indeed changing how businesses operated. Wal-Mart, for example, had combined the Japanese concept of Just In Time manufacturing with the Internet to implement real-time inventory control in its stores. The result: Wal-Mart stores had more floor space, less warehouse space, and a lot less idle inventory. Wal-

Mart could react faster than its competitors—in particular, Kmart—to changing consumer demands. It could move its goods quickly wherever they were wanted.

All of this caught the attention of John Garstka, a youngish, studious Air Force officer whose career had run into a bureaucratic cul-de-sac, thanks to the machinations of international politics and changing fashions of defense policy. Garstka was just two years out of the Air Force Academy in 1983 when he won a Hertz Fellowship, a prestigious program that sends students to graduate school to study how to apply hard sciences like physics, chemistry, and mathematics to real-world problems.

Garstka enrolled at Stanford to earn a degree in operations research. When he returned to the Air Force in 1985, the hot opportunity seemed to be the Strategic Defense Initiative, which Ronald Reagan had kicked off two years before. It was just the sort of challenge for an ambitious Air Force officer with credentials in both science and policy analysis, so Garstka went to work developing information systems that would control the space shield.

Alas, by 1989 the Cold War was winding down and defense budgets were starting to nose downward. Now Garstka's unusual combination of skills proved his undoing. The Air Force needed someone to work on personnel policies who had a background in information systems and operations research, plus clearances to special access programs. Garstka fit the profile.

So, when the plug got pulled on Star Wars, the Air Force yanked Garstka and his Stanford degree to Washington so he could, as he puts it, "sling computer code." It was not exactly the fast track to promotion, but while he was slinging, Garstka kept up a steady diet of *Business Week* to follow how the Internet and Just In Time were combining to change the way American companies did business.

It occurred to him that if a CEO could use Just In Time management methods and the Internet to reorganize his company's operations, why couldn't a military commander? At this point, a sort of Divine Intervention in behalf of national security occurred. Garstka happened to belong to the same parish as Arthur K. Cebrowski III, a Navy aviator who had flown more than a hundred missions over Vietnam and commanded an aircraft carrier during Desert Storm. As it happened, Cebrowski, like

Gartska, was a math major with a penchant for big theoretical concepts. They shared a common *Weltanschauung*. Garstka resigned his commission to become a civilian analyst at the Joint Staff, and worked off and on for Cebrowski during the next half-dozen years.[12]

Cebrowski was about to get his first star as a rear admiral and would specialize in developing requirements for the Defense Department's command and control systems. He needed an organizing concept, partly to make sense of all the different pieces of electronic hardware, partly to make sure they all worked together, and partly to explain to Congress and everyone else what the Defense Department was doing with the taxpayer's dollar. Garstka borrowed Scott McNealy's idea of network-centric computing and applied it to military operations. He showed the result to Cebrowski, who knew this was the idea that he had been looking for. They unveiled their creation, network-centric warfare, in an issue of the *Proceedings of the U.S. Naval Institute.*[13]

Network-centric warfare follows the basic idea of network-centric computing. It assumes that there is a worldwide grid of networked communications that any "platform"—ship, airplane, land vehicle, or just plain grunt—can plug into so that it can easily upload or download data. The effect is just like the Internet: what each platform happens to be is much less important than how they all work together.

Network-centric warfare called to mind the cartoon Peter Steiner published in *The New Yorker* just as the Internet boom was taking off. A black dog sitting at a computer keyboard leans over to tell a spotted canine, sagely, "On the Internet, nobody knows you're a dog." It was the same in network-centric warfare—it didn't matter whether you were a ship, airplane, land vehicle, or grunt. The most important idea in the whole concept was that the platforms were interchangeable, and that you could communicate with them all the same way.[14]

The idea caught on, first within the Navy, and then within the Defense Department as a whole, and even more so after Donald Rumsfeld became Secretary of Defense. Rumsfeld, Cebrowski, and Marshall were all part of the same mafia, or at least shared the same thinking. Network-centric warfare was written into Defense Planning Guidance—the Secretary's strategic plan, which is supposed to direct the services' own planning—and Cebrowski was put in charge of the Office of Force Transformation .

Network-centric warfare captured several strands of thought that had already taken root in the Defense Department in the 1990s. As we have seen, Andy Marshall had predicted that information technology was about to revolutionize warfare. The Joint Staff had also sponsored its own studies. The services were investing heavily in precision weapons. And, of course, when tech stocks were minting a millionaire a minute and televisions in the local bar were more likely to be tuned to CNBC than the NFL, it was hard for any defense official to avoid the question of how the Pentagon planned to change its own "business model" for the Information Age.

Network-centric warfare was this new model, and it represented a major change in thinking. Recall that the Army, Navy, and Air Force traditionally did not serve *à la carte*—they delivered units trained and equipped to execute operations according to their service's doctrine. Now just about everyone was beginning to realize that all military forces must work together better and adapt to changing situations faster. The solution depended largely on processing and—even more important—exchanging information more effectively. This was what network-centric warfare was all about.

The Defense Department will never be able to turn on a dime, but clearly something new is under way. Network-centric concepts are reshaping how U.S. forces fight. Critics who wonder whether the Defense Department is changing at all miss an important point. The thinking is changing faster than the hardware. U.S. commanders are using old forces in new ways. As we have already seen, in Afghanistan U.S. Marines operated far from shore. So did Navy SEAL special operations forces. The Navy also emptied an entire aircraft carrier of its fighter aircraft and bombers so it could be used as a floating base for Army helicopters and special operations forces.

This new thinking is having a radical effect on the planning and pacing of war. Traditionally, no commander would launch a bombing mission without meticulous planning. But, as Gartska observes, "In Enduring Freedom, 75 percent of the time aircraft did not even know what their targets were before they took off." Special operations forces on the ground would locate targets and pass the GPS coordinates to B-52 bombers orbiting overhead.[15] In other words, Just In Time military ops. Similarly, when the United States developed plans for military action against Iraq in 2003, U.S. commanders planned from the beginning an operation that melded Army, Navy, and Air Force units into a single, integrated network."

Also, critics who complain that defense transformation is moving too slowly forget that information technology is often cheap (at least in Defense Department terms), and is having a greater effect than dollar signs alone might suggest. Most of our military hardware still looks old—because it is—but the minor add-ons are making a big difference in how the U.S. military fights. Witness Afghanistan, for example, where U.S. forces combined old technology (B-52 bombers) and very old technology (Army Special Forces riding on horseback) with new information technology (laser designators, night-vision goggles, GPS, and networked communications).

The more important issue is not whether the U.S. military will adapt, but whether it will adapt faster and more effectively than our adversaries. So far, the evidence is mixed. Recall the October 1993 battle in Mogadishu, Somalia, that Mark Bowden made famous in his book *Black Hawk Down.*

American forces were in Mogadishu to support Operation Provide Comfort. The Somali national government had virtually vanished. Warlords competed for power, and they were using famine as a weapon to kill their opponents. The weapon of choice was the "technical," usually a Toyota pickup truck with a fifty-caliber machine gun mounted in back. Somali villagers fled their farms and poured into refugee camps. But the camps had no food, and the disaster was made even worse by a drought.

The United Nations agreed to send a relief mission, and the Americans were there to protect it. They soon became caught in the middle of the Somali civil war. The U.S. Army commander at the scene, Major General William Garrison, planned an operation to capture two Somali leaders who had been linked to groups that had fired on the Americans.

According to intelligence Garrison had received, the two Somalis had been spotted in a particularly bad Mogadishu neighborhood. Garrison decided to grab them. He planned to send a Delta Force team to assault their hideout, swooping in on stealthy Little Bird helicopters. Army Rangers would rope down from larger helicopters, Black Hawks, to form a protective perimeter. Then a column of regular Army troops in Humvees would move in to get everyone out.

"We knew we were going into what they called 'Indian territory.' We knew it wasn't a great part of town for us to be in," recalled Keni Thomas, one of the Rangers.[16] In fact, the Americans were going right into the middle of a Somali network, where the enemy had a clear information advantage.

The Somali commander, Mohamed Aidid, had set a trap. His objective was to shoot down a U.S. helicopter. Aidid knew that the Americans would not leave the crew behind, so the rescue would have the effect of freezing the Americans in place, making them a fixed target.

Aidid could not know ahead of time where his fighters might down the helicopter, but this was unimportant, because his forces were arranged as a network. Just about every man, woman, and child in Mogadishu seemed to be carrying an automatic weapon. And the Somalis had a simple but effective communications system: eyeballs, radios, and burning tires.

"I could see from a distance that they were dropping in from helicopters, so I immediately rushed the area," said Jamma Cabdulle, one of the militiamen who spotted the Rangers. This was tactical warning. When a Somali fighter shot down one helicopter with a rocket-propelled grenade, someone lit up a tire to mark the spot and signal everyone to converge. This was the command to converge and execute the operation.

As the Delta Force team and Rangers got bogged down, more and more Somalis converged on the scene. The enemy force grew exponentially. Soon just a few dozen Americans were facing thousands of Somalis. It was a perfectly executed swarming attack.

Several hours later the Americans rescued the trapped units with armored personnel carriers. American casualties totaled seventeen. The Somalis lost thousands. But Aidid achieved his objective. The sight of a dead American soldier being dragged through the streets of Mogadishu had exactly the effect the warlord wanted. Bill Clinton and his advisers decided that relieving Somalia was not worth the cost, and withdrew U.S. forces within a month.

The strike on the *Cole* was also a network attack. The Qaeda cells that almost sank the destroyer had been in Yemen for months and had scouted U.S. ships as they came through the harbor. So they knew when to expect their target and how to blend in with the harbor traffic. They also had effective, secure communications with each other. The *Cole*, on the other hand, was at the end of a long communications pipeline that stretched back through Central Command in Tampa. There were bits and pieces of information rattling around in U.S. government channels, but the information could not reach the people who needed it.

Al Qaeda cronies had appeared a few weeks before the attack on al-Jazeera and hinted that the terrorist organization planned to strike an American target soon. This, combined with intelligence warning, was enough to make the State Department evacuate nonessential staff from the U.S. embassy in Aden. Unfortunately, the *Cole* had no links to the State Department. The captain was unaware of the situation he was sailing into. So the safety of a $789 million destroyer depended totally on a handful of twenty-year-old sailors armed with unloaded M-16s, squinting into the noonday sun and trying to figure out why two guys on a skiff were waving at them as they approached.

Swarming even works in aerial combat. When the United States and Britain began to enforce no-fly zones over northern and southern Iraq during the 1990s, the Iraqis adopted a strategy that resembled the one Aidid used in Mogadishu. The Iraqi goal was to shoot down an American or British aircraft, hoping for a big political payoff. The Iraqi air defense system even used similar swarming tactics—spoofing our aircraft by directing their radars on them and sharing information throughout a sophisticated fiberoptic network when the aircraft returned fire or took evasive action.

In the Information Age, it's not just smart weapons that win wars. It's the total package—the total information picture—that is important. Even with long-range precision weapons, you still need a network that gets you inside your opponent's decision cycle. Aidid beat the Army in Mogadishu and Al Qaeda beat the Navy in Yemen because, in both cases, they had better information about us than we had about them—exactly the same situation that existed on September 11, 2001.

Information superiority works on many levels. It can be in the guidance system of a weapon. It can be in our ability to see the enemy. It can be whether we even know a particular enemy exists. About the only thing we can be sure of is that a smart opponent will try to learn what our weakest link is. It doesn't matter whether the strategy is swarming or zapping, information is decisive.

Chapter 12

KILLING

One of the byproducts of digital, dial-a-target precision warfare is that it makes bombs so precise, so accurate, that we can target individual people. In short, modern war raises an ancient issue: assassination.

There are few issues as controversial as assassination. The odd thing is these attitudes about assassination often seem to depend a lot on when you ask the question. Americans run hot or cold—*very* hot or *very* cold—on the issue.

In fact, this is exactly why the CIA found itself in so much trouble over assassination in the 1970s. In the 1950s and 1960s many Americans worried about the Soviet Union the same way they have worried about Al Qaeda, Iran, or Iraq in recent years. Not surprisingly, television shows like *Mission: Impossible* were big hits with viewers. Just about every plot had the Impossible Mission Force off in foreign countries breaking local laws, and often killing people, in the name of Cold War–era national security.

The CIA was simply doing exactly what Mr. Phelps was asked to do each week: "Should you or any member of your IM force be caught or killed, the Secretary will disavow any knowledge of your actions. Good luck, Jim."

Alas, a few years later, attitudes changed, and the CIA was caught in the whiplash. In the 1970s it seemed that no one thought—or had ever thought—assassination was acceptable. *How could you do such a thing?*

More recently, attitudes flipped back over. After the United States linked Osama bin Laden to the 1998 bombings of U.S. embassies in Kenya and Tanzania, we launched a cruise missile strike against one of his bases in Afghanistan. Former Clinton administration officials told reporters after the September 11 attack that the missile strike proved just how hard they had tried to get rid of bin Laden; it was only a matter of bad luck, they said, that they had failed.

There's nothing like a full-scale terrorist attack against the American homeland to get the old juices flowing. But you can be sure the pendulum will swing back, and people will find assassination as repugnant as they ever did. So we need to think rationally, not passionately, about the issue.

The problem today is that modern weapons are so accurate and modern intelligence and communications systems are so sophisticated that it often seems impossible *not* to target a particular person. To make matters even more complicated, our opponents today are often small terrorist cells, where the members may be individually known. The results are issues that can be incredibly complicated, especially in the heat of battle. For example, recall the exploits of the Predator in Afghanistan and the subsequent pursuit of Al Qaeda terrorists.

The Predator is an unmanned aerial vehicle, a UAV in military parlance. In effect, UAVs are big, sophisticated model airplanes. The Air Force originally developed the Predator in the mid-1990s for surveillance. Unlike a satellite, a UAV can linger over a single spot for hours. Unlike a manned aircraft, you don't have to send condolences to loved ones when a UAV gets shot down.

The basic idea for a UAV had been around for a long time. A cruise missile is essentially a UAV on a pre-set, one-way mission, and as we have seen, cruise missiles date back to Elmer Sperry's Aerial Torpedo. During the 1930s, the United States and Germany both developed remote-control ships and aircraft for use in target practice, and both used unpowered, remote-control "glide bombs" during World War II. After the war, the CIA developed a supersonic UAV called the D-21, and the Air Force also flew some early UAVs in Vietnam.

As in the case of precision weapons, the event that made UAVs practical was the Information Revolution. GPS made it easier to track a UAV from hundreds, even thousands, of miles away. Microchips and low-cost computers simplified the control problem. And lighter-weight sensors made it possible to collect better data.

Put it all together, and you have a floating eye-in-the-sky. It took a while for UAVs to prove themselves, but once they did, it did not take long for someone to figure out the next logical step: putting a small missile on the Predator. This transformed the UAV into an unmanned combat aerial vehicle (UCAV, pronounced "you-cav"). By February 2001 the Air Force had worked out the kinks. The CIA, which had also been operating Predators, began flying the armed version.

Now the United States had a unique new weapon: One could watch a target for hours with a Predator, and pipe the imagery to a local command post, the Pentagon—in principle, even top officials in the White House. They could decide when to order the operator to fire, all in one seamless process that was not much different from an international teleconference.[1]

This was exactly what U.S. officials wanted. After September 11, former Clinton administration officials told reporters that they had tried for years to capture or kill Osama bin Laden. The problem, they said, was getting "eyes on target"—actually seeing the man, tracking him, obtaining approval to strike, and then doing the deed. The armed Predator seemed to answer the mail.

But it did not take long to discover that, new technology or not, it was still a lot harder to find a specific person and kill him than one might think. As mentioned earlier, it was the CIA that got the armed Predator into action in Afghanistan first. The story, and the aftermath, hint at the controversies that lie ahead.

According to the account that later made the rounds, just after the United States began bombing Afghanistan one of the CIA's armed Predators spotted a convoy fleeing Kandahar. Because of the size of the convoy—100 or so people—the team operating the aircraft thought it might be carrying Mullah Omar, the leader of the Taliban regime.

However, the CIA did not have the authority to open fire. President Clinton had signed a so-called "lethal finding," permitting the CIA to take actions in countering terrorism that might cause the death of a suspected terrorist. President Bush had reaffirmed the CIA's authority. But *military* targets were not part of the authorization.

So the Predator operators passed the information to the Defense Department, where it ultimately reached Central Command (CENT-COM) headquarters in Tampa, Florida, where General Tommy Franks was directing the war. Meanwhile, the people in the convoy took cover in a building—possibly a mosque, possibly a residence.[2]

Franks reportedly said, "My JAG doesn't like this," referring to his judge advocate general, or legal counsel. The "lawyer in the loop" is a standard feature of modern command and control systems. CENTCOM ordered the Predator to shoot at a nearby target to flush out the occupants of the building. But the Taliban entourage, and Mullah Omar, got away.

Pundits had a field day. They claimed Franks muffed the call. Some even began a whispering campaign against him, claiming that President Bush wanted his head.

Yet the following February a very similar incident showed how a controversy can occur for exactly the opposite reasons. Another CIA Predator spotted fifteen to twenty people on the side of a hill near Zhawar Kili. By this time the fighting for the cities was over, and U.S. forces were looking for Al Qaeda cells that had taken to the country-side. One of the members of the group was a tall man with a beard. He seemed to be giving the other people orders. (Osama bin Laden stands about six feet four.)

By now U.S. officials had worked out a new "memorandum of understanding" that allowed the CIA greater latitude. As one official said, "The President has given the Agency the green light to do whatever is necessary. Lethal operations that were unthinkable pre–September 11 are now underway."[3] The Predator opened fire, and the missile raced toward its target.

Officials and critics alike later disagreed on exactly who was in the group, whether it was a bunch of locals gathering scrap metal or a band of fleeing Al Qaeda fighters. But after all the wreckage was examined and body tissues were analyzed, one thing was clear: The tall guy may have had an authoritative bearing, but he was not Osama bin Laden.

Nine months later the United States used the missile-toting Predator yet again. But this time, the target, Qaed Senyan al-Harthi, was thousands of miles from Afghanistan. Al-Harthi was riding in a car with five other passengers along a highway in Yemen. Al-Harthi had become one of the most sought-after members of Al Qaeda after the Taliban regime collapsed. Many American experts believed al-Harthi had been involved in planning the attack against the U.S.S. *Cole.*

Witnesses first thought the car simply exploded, possibly after hitting a mine. Later some claimed they had seen a helicopter flying nearby. The press began to investigate, and soon figured out that the "helicopter" was in fact a Predator, which they later linked to the CIA.

The Predator fired a missile, scored a direct hit, and the car carrying al-Harthi—apparently carrying fuel or ammunition—exploded. Nothing was left but scraps of metal and charred debris. U.S. officials made it clear they were pleased with the operation. The CIA officially had no comment.

Today the United States can put a bomb at a precise place at a precise time almost anywhere in the world, and often with top officials watching the action as it occurs. And the technology will only get better. Imagine UAVs that can orbit a building or shelter for weeks rather than hours, waiting for a suspect to appear. Imagine satellites in low earth orbit that can send a non-nuclear, precision-guided warhead to any spot on earth in fifteen minutes. (Treaties prohibit nuclear weapons in space, but conventional weapons are okay.)

Alas, we have not thought out the details. So, thanks to the Information Age, it's time to talk about assassination again.

Despite the intense emotions assassination raises, it rarely gets the kind of hard-headed, systematic analysis that we routinely devote to other national security issues, like defense budgets or U.S. membership in the World Trade Organization. That's too bad, because there are four basic questions we need to answer.

First, what exactly constitutes "assassination"? Second, does assassination break any constitutional principles, or treaties that the United States has agreed to? Third, does it work? And fourth, if we do carry out assassinations, who should do it?

Assassination is contentious partly because people on different sides of the issue are often arguing about different things.[4] Is killing during wartime assassination? Does assassination refer to killing people of high rank, or can anyone be the target of assassination? Does it matter if a member of the armed forces does the killing, a civilian government official, or a hired hand?

The term *assassin* comes to us from Arabic—*hashshashin,* or those who use hashish. Members of a tenth- to twelfth-century Islamic order were specialists in killing infidels—Crusaders—and were promised paradise if they died in action, a paradise that included, apparently, smoking dope.[5]

Today, if you look up the definition of *assassination*, the dictionary usually refers you to the verb *assassinate,* which is defined as "to injure or destroy unexpectedly and treacherously" or "murder by sudden or secret attack usually for impersonal reasons." In other words, assassination is murder—killing a person—while using secrecy or surprise. Assassination stands in contrast to murder without surprise (e.g., a duel). Also, assassination is not murder for personal gain or vengeance; assassinations support the goals of a government, group, or cause.

People often associate assassinations with prominent people, but strictly speaking, assassination knows no rank. Leaders are frequently the targets of state-sponsored assassination, but history shows that generals, common soldiers, big-time crime bosses, and low-level terrorists have all been targets too. Also, it does not seem to matter how you kill the target; it makes no difference if you use a bomb, a booby trap, a bullet, or a bazooka. As long as you target a particular person, it's assassination.

Surprisingly, there are no international laws that ban assassination. The closest thing to a prohibition is the 1973 Convention on the Prevention and Punishment of Crimes Against Internationally Protected Persons, Including Diplomatic Agents, or, as it is usually called, the Protected Persons Convention. This treaty (which the United States signed) bans attacks against heads of state while they conduct formal functions, against heads of government while traveling abroad, and against diplomats while performing their duties.

The Protected Persons Convention was supposed to make sure governments can negotiate even during war. Without it, countries might start a war (or get drawn into one) and then find themselves unable to stop because there was no leader at home to make the decision to do so and because their representatives were getting picked off on their way to the cease-fire negotiations.

But other than these narrow cases, the Protected Persons Convention says nothing about prohibiting assassination, and it applies only to officials representing bona fide governments and "international organizations of an intergovernmental character." So the convention shields the representatives of the United Nations, the World Trade Organization, and the International Red Cross. It does not protect bosses of international crime syndicates or the heads of terrorist groups such as Al Qaeda.

Another treaty that some might think would include an assassination ban is the Hague Convention with Respect to the Laws and Customs of War. The Hague Convention states that "the right of belligerents to adopt means of injuring the enemy is not unlimited." (This was a bold idea in 1907, when the convention was signed.)

Under the Hague Convention, combatants are entitled to the treaty's protections, but they are also obliged to obey its rules. For example, combatants must wear a "fixed distinctive emblem recognizable at a distance." Wear the emblem while fighting, and you are entitled to be treated as a POW if captured; fail to follow the dress code, and you might be shot as a mere bandit.*

The closest the Hague Convention comes to banning assassination in combat is when it prohibits signatories from killing enemy combatants "treacherously." But when the convention refers to "treachery," it is merely referring to fighting under false pretenses (e.g., flying the enemy's flag or wearing his uniform to lure him to his death).

The Hague Convention specifically permits "ruses of war." Snipers, buried mines, deception, camouflage, and other sneaky tactics are all okay. One might even argue that, since the convention (and related agreements, like the Geneva Convention) prohibits *indiscriminate* killing, assassination of combatants—the most precise

---

*When the United States imprisoned some 300 Taliban and Al Qaeda fighters at Guantánamo Bay, Cuba, U.S. officials argued that they were not entitled to Hague Convention protection because they did not wear uniforms or carry a "fixed, distinctive emblem." Wags suggested the fighters might have avoided this technicality if Taliban leaders had just bought factory-second Nike baseball caps and issued them to the troops. At least it might have reduced a war crime prosecution to an intellectual property litigation.

and deliberate killing of all—during war is exactly what the treaty calls for. As long as a country can justify self-defense, it can shoot, bomb, burn, or otherwise eliminate anyone in the military chain of command of its opponent.

As a result, the main legal constraints on assassination are rules nations impose on themselves. The U.S. government adopted such a ban in 1976, when President Ford, responding to the scandal that resulted when the press revealed CIA involvement in several assassinations, issued Executive Order 11905. This order prohibited "political assassination," and essentially reaffirmed an often-overlooked ban that Richard Helms had adopted for the CIA four years earlier when he was Director of Central Intelligence. Jimmy Carter also reaffirmed this ban in 1978, and Ronald Reagan went even further in 1981. His Executive Order 12333 banned "assassination" *in toto* and remains in effect today. Even so, there has been a disconnection between our policy and practice.

The United States has tried to kill foreign leaders on several occasions since 1976, usually as part of a larger military operation. In a 1986 air strike, U.S. warplanes bombed targets in Libya, including Muammar Qaddafi's tent, in retaliation for several terrorist attacks. During Desert Storm in 1991, we bombed Saddam Hussein's official residences and command bunkers. And, as we have seen, after the bombings of U.S. embassies in Kenya and Tanzania, the United States tried to intercept bin Laden at one of his bases in Afghanistan.

In each case, U.S. officials insisted that our forces were merely aiming at "command and control" nodes or a building linked to military operations or terrorist activities. In each case, however, the same officials admitted off the record that they would not have been upset if Qaddafi, Saddam, or bin Laden had been killed in the process.

The so-called "lethal findings," authorizing covert operations in which there is a good chance that an unfriendly foreign official might be killed, are another case in which the assassination ban has been fudged. For example, the CIA-backed covert operation to topple Saddam in 1996 probably would have killed him in the process, given the record of Iraqi leadership successions (no one has left office alive).[6]

In short, the unintended result of banning assassinations has been to make U.S. leaders perform verbal acrobatics to explain how they have tried to kill someone in an operation without really trying to kill him. One has to wonder about the wisdom of any policy that allows officials to do something, while requiring them to deny that they are doing it.

Whether assassination has been effective has depended much on what leaders have expected it to achieve. Most, but not all, attempts to change the course of large-scale political and diplomatic trends have failed. Assassination has been more effective in achieving small, specific goals.

Indeed, past U.S. assassination attempts have usually failed to achieve the minimal level of success: killing the intended target. According to the available information, *every* U.S. effort to kill a high-ranking national official outside a full-scale war has failed. This record is so poor that it would be hard to find an instrument of national policy that has been less successful in achieving its objectives than assassination.

Consider the record: The CIA supported assassins trying to kill Patrice Lumumba of the Congo in 1961 and repeatedly tried to assassinate Fidel Castro between 1961 and 1963.[7] And, as noted, in recent years the United States has tried to do away with Qaddafi, Saddam, and bin Laden.

This record is remarkably free of success. Castro, Qaddafi, Saddam, and bin Laden all survived the assassination attempts listed above. Moreover, they were not cowed. Qaddafi continued to support terrorism (e.g., the bombing of Pan Am Flight 103). Saddam managed to outlast the terms of two Presidents who wanted to eliminate him (George Bush and Bill Clinton), while also continuing to support terrorism—and develop weapons of mass destruction.[*]

One might have predicted this dismal record just by considering why American leaders have resorted to the assassination option. More often than not, assassination is the option of choice when nothing seems to work but when officials think that they need to do *something*.

[*] As noted, the United States did manage to nail al-Harthi in Yemen. However, as of this writing (early 2003), Osama bin Laden and Mullah Omar are apparently alive. More to the point, hardly anyone believes that killing *any* specific person would eliminate Al Qaeda.

When diplomacy is ineffective and war seems too costly, assassination becomes the fallback, without any clear understanding of its effectiveness or ramifications.

True, some other countries have been more successful, in that they have killed their target. For example, after the terrorist attack on Israeli athletes in the 1972 Munich Olympics, Israeli special services tracked down and killed each of the Palestinian guerrillas who took part (they also killed an innocent Palestinian in a case of mistaken identify). In 1988 Israeli commandos killed Khalil al-Wazir, a lieutenant of Yasir Arafat's, in a raid on PLO headquarters in Tunisia. More recently, Israel has killed specifically targeted Palestinian terrorist leaders—for example, Yechya Ayyash, who was killed with a booby-trapped cell phone.

Other countries have also attempted assassinations with some degree of tactical success. During the Cold War, the KGB and other communist bloc intelligence organizations were linked to several assassinations. Most recently, the Taliban regime in Afghanistan was involved in the assassination of Ahmed Shah Massoud, the leader of the Northern Alliance opposition.

Yet even "successful" assassinations have often left the sponsor worse off, not better. U.S. officials had hoped that the removal of Ngo Dinh Diem would produce a South Vietnamese government that enjoyed more popular support. But when a group of South Vietnamese generals succeeded in eliminating him, the resulting government was no more popular or effective, and the United States was sucked even deeper into a misconceived war.

Similarly, German retribution against Czech civilians after the 1942 assassination of Nazi prefect Reinhard Heydrich by British-sponsored resistance fighters was especially brutal.[8] The 1948 assassination of Mohandas Gandhi by Hindu extremists led to violence that resulted in the partition of India. The assassination of Abraham Lincoln (carried out by a conspiracy some believe was linked to the Confederate secret service) resulted in Reconstruction.[9]

Some people argue that the assassination of Yitzhak Rabin by a Zionist extremist in 1995 may be the exception—it derailed the Middle East peace process. But one can just as easily argue that Ehud Barak was even more willing to make concessions for peace than

Rabin, and that Yasir Arafat would have derailed the peace process later anyway—as he did. In short, assassination has rarely been reliable in shaping large-scale political trends the way the perpetrators intended.

When assassination accomplishes anything useful, it is usually by depriving an enemy of the talents of some uniquely skilled individual. For example, in 1943 U.S. warplanes shot down an aircraft known to be carrying Admiral Isoroku Yamamoto, the architect of Japan's early victories in the Pacific. His loss hurt the Japanese war effort.

The problem is, picking off a talented individual is almost always harder than it looks. One paradox of modern warfare is that despite modern weapons, sensors, and communications systems, it is still difficult to kill a particular person. One has to know exactly where the target will be at a precise moment. This is almost always hard, and it is especially hard when the target knows he is being pursued.

The more complex issue is not whether assassination works, but whether we should resort to it. The morality of assassination depends mainly on the question of whether and when one can justify state-sponsored killing.

Although Americans do not like the idea of their government killing people, they do not completely reject it either. Most would agree that their government should be allowed to kill in at least two situations: policing and going to war. In addition, many Americans believe the government should be allowed to impose capital punishment for at least some crimes.

But, while almost everyone agrees that the police should be able to protect themselves and others when making an arrest, few would want police to run down their suspects with the expectation that they would routinely kill them. Similarly, even the staunchest supporters of capital punishment believe that an accused criminal is entitled to a fair trial. All of this would suggest that assassination is unacceptable as a substitute for situations that really call for law enforcement.

The only time we should consider assassination is when we do need to protect ourselves from a clear, immediate, lethal threat from abroad. In other words, assassination is a military option—that is, an act of armed self-defense, which is to say, war. We need to understand that assassination is warfare, and when we try to assassinate, we are not avoiding war, we are going to war, with all the risks and costs that war brings. These include diplomatic consequences, the danger of escalation, the threat of retaliation against our own leaders, and possible retaliation against American civilians.

But if we accept assassination as an acceptable (or unavoidable) option, who should do it?

Nine days after the attacks on New York City and Washington, President George W. Bush addressed Congress and the nation in a televised speech. He wanted to explain his plan for responding to the attacks. He also wanted to reassure a country that was shocked, scared, and angry.

Bush told the country to prepare for a new kind of fight. As he put it, "Americans should not expect one battle, but a lengthy campaign unlike any other we have ever seen. It may include dramatic strikes visible on TV and covert operations secret even in success."[10]

Eight months later, the Taliban regime had collapsed in Afghanistan. Forces from the United States and Britain were pursuing Al Qaeda fighters in the mountains along Afghanistan's frontier with Pakistan. Other forces were looking for Al Qaeda fighters in Pakistan, the Philippines, Yemen, and the sea lanes of the Indian Ocean.

Bush gave another policy speech, this time at the U.S. Military Academy at West Point, in upstate New York. Breaking the ice, he observed, "A few of you have followed in the path of the perfect West Point graduate, Robert E. Lee, who never received a single demerit in four years. Some of you followed in the path of the imperfect graduate, Ulysses S. Grant, who had his fair share of demerits, and said the happiest day of his life was 'the day I left West Point.'"

"During my college years I guess you could say I was"—Bush paused for effect—"a Grant man." The crowd laughed, thoroughly familiar with Bush's reputation at Yale.

Then Bush turned serious, and proceeded to define a policy that was even more aggressive. The President said, "The war on terror will not be won on the defensive. We must take the battle to the enemy, disrupt his plans, and confront the worst threats before they emerge." Continuing, he said, "Our security will require all Americans to be forward-looking and resolute, to be ready for preemptive action when necessary to defend our liberty and to defend our lives."[11]

Add the two speeches together, and you have a military policy that is a complete break with the past half-century. Cold War policy was based on deterrence: build highly visible forces, be prepared to retaliate, and leave the opponent too fearful to attack. The new policy proposed hidden forces, shooting first to root out, pinpoint, and destroy the enemy before it attacks.

There is an unavoidable logic of modern war: Combat has become mainly a race to reach the end of one's decision cycle first. Whoever wins that race gets to take the first shot. And because modern weapons are so precise and deadly, whoever wins the opportunity to take that first shot wins, period. This is a formula that argues for a policy of preemption and, given the realities of modern military technology, it means possibly attacking individuals in the process.

Rarely, if ever, had a leader so openly declared a policy where he committed his country to covert action and striking first. So what does this really mean? And what issues are raised by the intersections of covert action, striking first, and killing specific individuals?

The legal definition of covert action appears in statute—specifically, Title 50, Chapter 15, Section 413b of the U.S. Code. Covert action is any "activity or activities of the United States Government to influence political, economic, or military conditions abroad, where it is intended that the role of the United States Government will not be apparent or acknowledged publicly." In other words, covert operations are *deniable activities*.

It is important to understand that covert operations are not *secret activities*, which are supposed to be concealed completely. Some high-tech weapons, for example, are secret. Hiding them prevents enemies from developing countermeasures. Some diplomacy is secret too, so U.S. officials have greater flexibility in floating ideas or negotiating the early part of an agreement.

Most covert operations are, in fact, entirely visible except for their connection with the U.S. government. Indeed, almost anything the

United States has done as a covert action—paramilitary operations, security assistance, etc.—has been done overtly on other occasions. That is why the first question to ask about any proposed covert operation is "Why do you want to do it covertly?" *Why is concealing the role of the U.S. government essential to its success?*

Usually, the only good reason for covertness is that open knowledge of U.S. responsibility would make the operation impossible. For instance, some propaganda we aim abroad will have a greater impact if it seems to come from a neutral source. Payoffs to a foreign official could give his rivals political ammunition if they became public. That was also why the United States has used covert forces to help governments like Pakistan and the Philippines track down Al Qaeda cells.[12]

But killing people is different. The methods of terrorists and armies are hard enough to distinguish today. Both are using the same tactics—dispersed networks of cells that hide until it is time to strike. Both increasingly have access to the same technologies.

Today the main differences between terrorists and armies are the rules they follow. Soldiers wear uniforms because governments are supposed to take responsibility for their actions when they exercise their legitimate right of self-defense. Armies train to avoid killing noncombatants, and distinguish themselves from civilians to avoid sucking innocents into combat.

Terrorists do just the opposite. They target civilians intentionally to create fear and confusion; that is, after all, the definition of terror. Terrorists hide their identities to be harder to find, to blend into the rest of the population, and make it harder to fight them without harming noncombatants.

Killing covertly would undercut our standing as a country trying to uphold the rule of law. We *don't* want terrorists like bin Laden and al-Harthi, dictators like Saddam Hussein, or other evildoers to "just disappear." If someone plans to pull a trigger, release a bomb, fire a missile, or stick a knife in the side of someone's neck, it should not be a covert operation. When we need to target individuals, it should be like any other military action. We should use U.S. forces reporting through a clear, public chain of command. If we kill covertly—that is, if we hide U.S. responsibility—our operations begin to resemble those of the terrorists.

The United States did not ask for the threats we currently face, and killing in behalf of the state will always be the most controversial policy issue of all. That is why we need to use blunt language and appreciate exactly what we are proposing. Sugarcoating the topic only hides the tough issues we need to decide as a country. If we need to target specific people for military attack, it is important that we get it right.

If we do not want to kill individuals covertly, what other choices do we have? There is another kind of military operation that will likely become more important in the Information Age, when stealth and getting ahead of the decision cycle of your opponent are critical. This option is "direct action." It has received much less attention than covert action in the past, but we will likely hear more about it in the future.

The Defense Department's *Dictionary of Military and Associated Terms* defines direct action as "short-duration strikes and other small-scale offensive actions by special operations forces or special operation–capable units to seize, destroy, capture, recover, or inflict damage on designated personnel or matériel." Strip away the jargon and you are talking about using troops to ambush terrorist groups, raid weapons shipments in transit, and rescue hostages.

One reason why direct action will become more important is to avoid the kinds of mistakes that occurred with the Predators in Afghanistan. Sometimes, to make sure you are targeting the right person and know you hit him, there is no substitute for having people at the scene. Also, terrorists will learn from the mistakes Al Qaeda committed by becoming too visible in Afghanistan. They will adopt a lower profile, and take refuge in countries that are not quite hostile to the United States but unwilling to allow us to base military forces in their territory. In many cases, the only solution will be small forces, striking quickly.

In direct action, soldiers wear insignia when they fight. This is the important distinction between covert operations and direct action. It is a key difference between using innovative military tactics to eliminate terrorists, rather than acting like terrorists to eliminate terrorists. Direct action complies with international law. Also, when a direct action goes sour, politicians have no place to hide. Keeping military action overt forces officials to decide whether they can justify going to war, and whether they want to take that step. In other words, direct action, unlike covert action, keeps officials under public scrutiny.

Unfortunately, the U.S. military is inadequately prepared for this new kind of combat. Deploying even a small force today is a big operation. Putting almost any combat unit into the field usually requires several days—even weeks—of preparation. Few military units are trained to react quickly, and those that could get to the scene quickly would have little logistics support when they arrive.

This is why we have been using the CIA as a quick-response military unit. Yet using the CIA for combat operations will likely hurt us in the long run. Aside from the fact that covertness blurs the link between officials and the military operations they approve, few organizations can do more than one thing well. If the CIA prepares to fight wars, it will be less prepared to conduct espionage. The last thing you want to do if you are trying to maintain safe houses and dead drops is to call attention to yourself with helicopters, armoured vehicles, and guys in fatigues carrying automatic weapons. Recall, for example, how much has been reported about the CIA's activities in Afghanistan, or about the CIA's role in flying Predator operations.

For the new wars, U.S. military forces need to develop their own capabilities to "get there first"—small, highly mobile combat forces, combined with the infrastructure they need for logistics, communications, and supplies. Some military activities—especially preparations and support—will need to be secret. But lethal combat should always be public and attributable. Let the spies be spies and the soldiers be soldiers. We will need them both.

Chapter 13

AN ELECTRONIC PEARL HARBOR?

By now it ought to be clear how much our national security depends on our policies for promoting, regulating, and controlling information technology. We can't defeat terrorist groups, rogue states, and other threats unless we can stay at least one step ahead of their decision cycle. That often depends on whether we can crack into their networks—and protect our own.

Type the phrase "electronic Pearl Harbor" into an Internet search engine like Google or Alta Vista, and you will find a surprisingly large number of hits. In the winter of 2003 Google would turn up 2,390 entries on the World Wide Web. The doomsday scenario du jour varies, and often seems linked to the latest headlines more than anything else.

During the stock market boom of the late 1990s, experts warned that terrorists might crash the U.S. economy by hacking NASDAQ's computers. When there are an unusual number of airline accidents, they warn of hackers bringing down the air-traffic control system. During the California energy crunch of 2000, some pundits warned that terrorists might hijack computer systems at electric plants and shut down the generators. "You can black out whole cities," warned Anjan Bose, a professor of power engineering at Washington State University.[1]

Sometimes I blame Matthew Broderick for all of this. In the 1983 film *WarGames,* Broderick plays David Lightman, a teenage geek who gets his kicks by breaking into the computer systems at video game developers. Lightman stumbles into a computer at a defense contractor, and somehow manages to get into the warning network operated by the North American Aerospace Defense Command (NORAD). Lightman comes close to triggering a missile strike against the Soviet Union. But there is a happy ending. The computer figures out that there are no winners in a nuclear war. Lightman even gets the girl, Jennifer Mack, played by Ally Sheedy.

The plot was perfect for the 1980s, when people worried that the Soviet Union and United States might start lobbing nuclear weapons at each other. If anyone raised the issue of a catastrophic computer malfunction, naturally the risk that first popped into people's minds was that the malfunction might trigger thermonuclear Armageddon.

These images can be hard to shake. A few years after *WarGames* hit the screen, I was visiting the NORAD command center at Cheyenne Mountain in Colorado. Someone jokingly mentioned the film to the colonel hosting us. He just rolled his eyes. They were still trying to re-assure people that it couldn't really happen.

Network warfare will hinge largely on who can attack and defend communications and computer systems. But to deal with the issue effectively, we always need to focus on the central question—how could someone use such an attack? As we will see, even top officials are often preoccupied with hypothetical, abstract threats, and insufficiently concerned with plausible, immediate ones.

Once TS3600.1 was in place and the new Clinton administration appointees began to settle into office in 1993, organizations throughout the Defense Department began to look at information warfare more closely. Some officials asked, how can you use information warfare to win a war? And, if you took information warfare to its logical extreme, what would the result look like?

When Major General Frank B. Horton III arrived at the Pentagon in 1993, he was a civilian, for the first time in three decades. Barry Horton was a former Eagle Scout who had spent most of his Air Force career specializing in intelligence and strategic nuclear forces. Along the way, he also found time to earn a doctorate from Harvard in political science.

Horton had met William Perry, a longtime member of the Democratic Party's defense policy brain trust, while the two were attending a conference some time back. When Bill Clinton asked Perry to become his Deputy Secretary of Defense, Perry invited Horton to come on board. Horton assumed the title of Principal Deputy Assistant Secretary of Defense (C3I). His boss, Emmett Paige, gave him carte blanche to pursue whatever issues he thought were important. Horton decided fto focus on information warfare. The only drawback for Horton was that he had to leave his military career behind. Since Horton was eligible for retirement, he left the Air Force after thirty years and arrived at the Pentagon as Dr. Horton, sans uniform.

Horton's interest in information warfare was another example of how twenty-first-century information warfare is rooted in twentieth-century planning for nuclear war with the Soviet Union. In his last assignment to the Strategic Air Command, Horton had been in charge of the target list used for the Single Integrated Operations Plan (SIOP), or U.S. plan for strategic nuclear war. Part of the SIOP was attacking the Soviet command and control system, the node analysis problem that Ron Knecht and others had worked on.

Horton also got drawn into the "blue" side of the problem—protecting U.S. command and control. The main goal is ensuring that U.S. leaders can always get a "go" signal to our missiles and bombers, while at the same time making sure that no one launches a nuclear strike without proper authorization. There was always the lingering concern that General Jack D. Ripper might suddenly decide one evening to settle the Cold War once and for all with a Dr. Strangelove solution.

Insiders still refer to the Horton Report, a study Horton did in 1985 that served as the basis for the Strategic Air Command's solution. Horton came to appreciate the importance of information systems to military operations, both how things can go wrong if your information systems do not work as designed, and how you can attack your opponent's own system.

All of this put Horton in a position to pick up where Duane Andrews, Paul Strassmann, Ron Knecht, and Bob Carpenter had left off. Horton knew that the problem was not just the vulnerability of U.S. military communications and computers. Civilian systems were an even bigger problem.

The vast majority of military communications—nine-tenths or more—travel over commercial links. Defending military networks is virtually inseparable from defending commercial networks. If an attacker takes out key commercial links, U.S. armed forces might be unable to operate. Add the fact that the Pentagon buys most of its hardware and software off the shelf, and everyone soon understood that the private sector was a weak link in national defense.

This problem had become more complicated in 1984, when Judge Harold Green broke up the Bell System monopoly. Now the Defense Department had to deal with several companies and operators to protect the American telecommunications system, whereas before it only had to deal with one. When the Internet took off—and the government decided that the Information Superhighway would be privatized—the number of companies with a role in protecting the links multiplied into the hundreds, even thousands.

The question was, how could the Defense Department get the attention of the private sector? For that matter, how do you get the attention of top defense officials who had more pressing matters than some esoteric topic the technoids were calling "information warfare"?

Horton came up with the idea of a game simulating what a cyber attack might look like—or, as one player said, "raising the issue's 'Holy shit!' content." Horton recalled some work by Roger Molander and Peter Wilson, two RAND analysts who had designed games simulating situations that threatened to trigger a nuclear war.

Molander and Wilson designed their games as if they were three-act plays. The trick was that the players joined the action at the halfway point—Act II, after the action was underway but before the bombs went off. Depending on how the players dealt with the crisis, in Act III they either defused the crisis or blew up the world in a nuclear exchange. The players would then go back to Act I to see how they might have laid the groundwork for a better outcome. This, of course, was the whole point of the exercise—laying better groundwork today for a better outcome tomorrow. Molander and Wilson accordingly called their three-act plays "Day After . . ." exercises.

Horton asked Molander and Wilson whether they could conjure up situations in which an adversary threatened not a nuclear strike, but an attack on American computers and communications systems. To make things interesting, the attack would come in the middle of some other crisis, like a war in the Middle East or a Chinese strike against Taiwan.

Soon the Saturday morning "Day After . . . in Cyberspace" games were a hot ticket in Washington. Molander and Wilson would get twenty or thirty officials together in a room at RAND's old Washington office on M Street. Everyone would receive their notebooks and role assignments a couple of days before ("Waddaya mean, I'm at the State Department?") The participants spent a few evenings reading their situation summaries and briefing materials, and showed up ready to play.[2]

In a typical game, Iran might lead a faction of OPEC hawks to hike petroleum prices to $60 a barrel. Saudi Arabia would resist, and a summit meeting of oil ministers would break up in a shouting match. Iranian troops mobilized, and U.S. Navy forces went to aid the Saudis.

Then mysterious things started happening—control systems at oil refineries near Dhahran went haywire, causing the plant to blow up. Amtrak Metroliners somehow got switched to the wrong tracks and slammed into freight trains heading in the opposite direction. Power grids went down in the Northeast, and Pentagon logistics offices detected sophisticated hackers probing their computers from somewhere in the Middle East.

And so amid the Styrofoam coffee cups and Coke cans in the RAND conference room, the players planned and executed the next regional war, with a cyber conflict taking place around the edges. Everyone had a lot of fun—and got worried. Defending against cyber attacks was something new, and U.S. officials discovered they were not sure how to deal with it.

The biggest problem was simply that you often did not know who the cyber attacker was. Sometimes you could not even tell an attack was underway. Did someone intentionally crash a computer to keep you confused? Or was it a normal system glitch? After all, everyone knew that computers were always crashing of their own accord. And, even if you think someone is screwing around with your computers, how do you respond? As one participating admiral said, "What am I supposed to do, nuke them for turning off our TVs?"

The games had the desired effect. The buzz in the Pentagon grew, and eventually word seeped out into the media. Now information warfare was not just an abstraction; it had a story line.

And that was when the whole issue got bent totally out of shape. If anything, Horton, Molander, and Wilson were too successful. Once the public started talking about information warfare, it seemed like every twenty-three-year-old male who was still living with his parents and had a computer in his bedroom imagined himself as a potential Luke Skywalker doing battle against the forces of darkness.

During the Internet boom people began staring into computer screens six or seven hours a day. They found it easy to imagine that some nefarious ninja info-warrior might reach right into the innards of their PC and grab their Quicken files or that nude image of Teri Hatcher they had downloaded from the Net. And, if the forces of darkness could do that, then surely they could crash any computer, anywhere in the world, and cause all sorts of mayhem.

Molander and Wilson would wince when they heard it, but quickly the catchphrase was on the lips of everyone who worked in the national security game: "electronic Pearl Harbor." Literally, it meant that some country or terrorist might attack U.S. computers in one sudden, bolt-out-of-the-blue strike, causing death, destruction, and mayhem. Armies would be unable to leave their barracks. Airliners would fall out of the skies. Wall Street would collapse.

Soon it seemed everyone in the Defense Department was talking about information warfare. Some even claimed that the Pentagon's PowerPoint warriors had programmed their computers to insert the term *electronic Pearl Harbor* into their briefings with a single keystroke.

Peter Wilson later tried to explain that the games were supposed to make people think, not predict the future. Cyberwar is a "weapon of mass disruption, not a weapon of mass destruction," he said. Of course it would be harder for American forces to fight if our computers crashed all at once at an inopportune moment. Certainly the confusion would make it tougher for leaders to make decisions. But kill people and hobble an economy by moving electrons around? No.

But the idea took hold. For example, James Adams, an otherwise respected author, published an article for *Foreign Affairs,* the otherwise esteemed journal of record for the foreign policy community. Adams described a 1997 Defense Department exercise called Eligible Receiver. According to Adams, "Red teams" from the National Security Agency (NSA) had simulated an enemy attack on U.S. computer systems using hacking tools they found on the Internet. "Over the next two weeks," wrote Adams, "the teams used the commercial computers and hacking programs they downloaded from the Internet to simultaneously break into the power grids of nine American cities and crack their 911 emergency systems."

Other writers have written similar accounts of Eligible Receiver, and even many government officials cite it as proof of the possibility for an electronic doomsday scenario. But according to William Crowell, this is not what Eligible Receiver was about. Crowell should know, as he was the Deputy Director of NSA in 1997 and oversaw the planning of the exercise.[3]

Crowell stresses that Eligible Receiver was aimed at checking the security of *Defense Department* computer systems. Officials had worried about their vulnerability at least since Dick Cheney approved the first directive on Information Warfare back in 1992. What made Eligible Receiver unique was simply that it was a large-scale "no notice" exercise. That is, it was the first live fire test.

The targets in the exercise were Air Force, Navy, and OSD computer systems (the Army declined to play). The NSA hackers who probed the Defense Department for vulnerable computers did what a good computer security consultant would do, except on a much larger scale—and with a better script, written by people who analyzed foreign military organizations and cracked into communications for a living.

One of the biggest challenges in staging the exercise was simply getting all of the required approvals up and down the chain of command. There were all kinds of legal issues for the lawyers to solve. Defense Department employees are protected against surveillance by the same laws that keep NSA from spying on other Americans.

Once the exercise was approved and carried out, the result that most concerned Defense Department officials was not just that their networks were vulnerable or that mayhem would result. What really concerned them was how quickly a sophisticated attacker could reconnoiter their networks, and how hard it would be to know that anything had happened, or who did it.

During the first week of the exercise, the NSA hackers used the agency's own Internet addresses. The officials staging the exercise had not quite completed the legal work. In particular, they had not yet received approval to use bogus Internet addresses; the law not only forbids NSA from monitoring Americans, it also restricts NSA from masking its identity. So the hackers went ahead, using their NSA accounts. The crews monitoring the hacking quickly figured out who was at work.

But when the hackers were later allowed to use an anonymous Internet address, no one was able to identify them or what they were doing. Indeed, the hackers had to leave behind calling cards—files they could plant and later identify—so that after the exercise was over, they could prove they had really cracked into the system.

As for hacking the commercial power grid, Adams claimed, "This exercise proved that genuine hackers with malicious intent could, with a couple of keystrokes, have turned off these cities' power and prevented the local emergency services from responding to the crises."[4] Not so. Getting approval just to attack Defense Department computers was hard enough. The exercise took two weeks; the approval process took about a year. Screwing around with Pacific Bell or ConEd was never on the table. The closest NSA ever came to a civilian system was explaining how they might have gone about the operation.

But suppose a hacker had been able to break into the corporate computer network of a power company, jump over into the systems controlling a generator, and shut down the system. What would he have accomplished? Adams never bothered to say whether, for instance, the generating plants might be able to reboot their computers, whether other utility companies might have immediately provided backup capacity, or, for that matter, whether any of the attacks would have caused any real damage.

The overreaction to the threat of a massive, all-out cyber attack has only gotten worse since. Consider Richard Clarke, a White House staff member responsible for cyber-security policy.

"We could wake one morning," Clarke told a group of telecommunications executives in 1998, "and find a city, or a sector of the country, or the whole country having an electric power problem, a transportation problem or a telecommunication problem because there was a surprise attack using information warfare." He warned, "There's a giant tsunami about to crash down behind us."[5]

Take cover. Or, better yet, get a grip. By painting doomsday scenarios, government officials lose credibility and, over time, their ability to influence the public. In doing so, they simply make it harder to build support for dealing with real threats.

Computer viruses are the classic example. Viruses probably get more publicity than any other computer security problem. Thanks to the Internet, when one gets loose, it can affect millions of people around the world in just a matter of hours.

Computer mavens usually give Fred Cohen credit for coining the term "computer virus" in a technical paper he published in 1984 while a grad student. Cohen, now at Sandia Labs, described a program that could infect someone else's software and cause it to operate in ways the original writer never intended.[6] But the basic idea for a self-replicating, and possibly mischief-causing, bug goes back to the earliest days of computing. John von Neumann, one of the first people to think about programmable computers, also postulated the idea of a program that could read and replicate itself. Later some technicians at Bell Labs actually wrote such a program.

Then, in the 1970s, computer scientists were trying to figure out how to move programs and data automatically from computer to computer. This is handy if, for example, you want to move the entry for an airliner from one air-traffic control computer to another as the aircraft flies coast to coast. Or you might want to divide a data set and move it through a network so that several computers can split up a big computation.

To do this, the computer scientists used the idea of a self-replicating program to invent what they called a "worm." The worm sends out a message from the computer on which it resides, asking the whereabouts of other computers on a network. When it gets an answer, it sends out a return message containing a copy of itself. The whole process is just supposed to lash several computers together, and worms were technical curiosities until December 1988, when Robert Morris, a Cornell grad student, released one on the Internet.

The Morris worm had a bug that was apt to crash any computer that received it. So as the worm moved through the Internet, it brought down one machine after another. The worm managed to crash 10 percent of all the computers connected to the Internet.

The press couldn't resist the story. The Internet was just becoming well known, and it was a typical "follies of technical hubris" yarn. Ever since, computer attacks have been a news staple, and especially in the late 1990s, when the World Wide Web took off and more people were getting online. Any incident was bound to attract more attention.

Recent incidents have usually involved viruses rather than worms. A virus is similar to a worm, but while a worm replicates itself to move from machine to machine, a virus makes copies of itself and leaves one on as many machines as it can infect. The ability of viruses to spread from computer to computer and do damage depends on how well they can take advantage of common features found in computers.

For example, all computers that use the Windows operating system—95 percent of all personal computers—use the same naming system to organize and recognize files. So if you can somehow sneak a program onto a computer, you can design it to look for the kind of file Windows uses to, say, save e-mail address books. If you include a routine that sends a copy of the program to each address, you've created a workable virus.

That's exactly what the Love Letter virus did when it appeared in May 2000. The virus was an e-mail attachment that looked like a love letter from a secret admirer. If you were foolish enough to open the attachment (or desperate enough for a secret admirer), the virus activated—reading your e-mail address book, replicating itself, and then sending a new message to everyone listed.

The biggest problem for the virus writer is to find some way to get the program on a computer. One way is to make the virus look like some other kind of program or file, and then trick the user into activating it. In the case of the Love Letter, the writer knew that a lot of people just would not be able to resist seeing what was inside the attachment, any more than a schmo who gets a trick can of "peanut brittle" can't resist opening it. The result is the same in both cases: exploding snakes spring out, making the victim look like a gull.

Investigators later traced the virus back to Onel de Guzman, a student in the Philippines who claimed he accidentally put the program on the Internet. Because the Internet goes just about everywhere these days and lots of people correspond with people overseas, the virus quickly clogged computers and server connections worldwide. In just three days it infected some 45 million computer users in over twenty countries.

More recent viruses have been of a variety called "distributed denial of services" (DDOS). This kind of virus is designed to scan the Internet for computers that have inadequate firewalls—the software programs or devices that are supposed to protect computers from viruses and intruders. DDOS viruses infect the vulnerable computers and lie waiting. Then, at a pre-set moment, the viruses activate, turning their host computers into zombies programmed to send mass volumes of junk messages to a targeted victim. (The target of the Code Red virus that appeared in late 2001 was the White House Web server.)

The difficulty in rooting out a DDOS incident is that any computer connected to the Internet is a potential zombie and a potential target. It is hard to alert every computer operator that might be connected to the Internet to inspect their system and eliminate the virus if they find it.

On the other hand, fixing the problem is not that hard either. Since all the attacking viruses were clones of the original, they all included common features that could be recognized, allowing their potential victims and other operators to filter them out. The creators of the viruses could not adjust their aim once they pulled the trigger. (There was an additional problem in that the virus contained several subroutines that would modify the code and restart the attack on a later date, but once a programmer had a specimen, he could look for these tricks too).

Some computer security experts say that much of the problem with viruses is simply that personal computers are the most often used device for connecting to the Internet, and PCs were never designed for that purpose. Personal computers were originally intended in the 1970s and 1980s to operate as stand-alone machines, so it never occurred to anyone to design safeguards into their software to protect them from intruders. A decade or so later, when everyone started going online, it was as though someone tore the Bubble Boy out of his protective suit and tossed him into a New York City street to fend for himself in a world of crime and disease. Ever since, we have been playing catch-up, trying to fix all the built-in flaws and features that make PCs so vulnerable.

But is this information warfare? Only by the standards of that twenty-three-year-old geek who still lives with his parents. It is hard to predict where a virus might wind up, whether it will land on the targeted computers, or whether it will go off at the intended time. No military officer would ever stake the lives under his or her command on such a scattershot approach. It is not a useful weapon.

Probably the absolute worst damage a virus could do is bog down the Internet for a while and make some sites unreachable because of the traffic jam. Meanwhile, users would find some alternative form of communication. (Ever notice that people always use the telephone to tell their Internet service provider that their system crashed; no one ever seems to use e-mail to tell the telephone company that their phone does not work.) There is so much redundancy in the nation's communications networks, and people know how to adapt, so it is hard to think of how a computer virus could bring the country to its knees.

But what about the cost of these virus "attacks"? The Morris worm incident not only brought the problem of viruses to the public's attention; it also began a tradition: hyping the estimated cost of virus episodes. *Wired* claimed that the Morris worm caused $15 million worth of damage, based on the time required to debug infected machines.[7] A decade later *The New York Times* reported that the Love Letter virus caused $15 *billion* in damage.[8]

If you believed these reports, the carnage and devastation of the Love Letter attack was in the same league as the most costly natural disaster in U.S. history, when Hurricane Andrew swept across Florida and the Gulf Coast (losses: $19 billion from 750,000 documented insurance claims, plus the loss of twenty-six lives—and the destruction of Homestead Air Force Base and an F-16 fighter).[9] Or, to put it another way, it would have taken as few as two Love Letter incidents to equal all the September 11 damage to New York (losses: $30–70 billion resulting from the collapse of three skyscrapers and destruction of several smaller buildings).[10]

It's hard to take these estimates seriously. The costs of virus attacks that the press reports are based mainly on estimates of hypothetical lost business and the time systems operators spend to reboot servers—as though the operators would otherwise be staying home.

Even top government officials often hype the costs of virus "attacks." For instance, in February 2002 Richard Clarke briefed a Senate subcommittee. The chairman, Charles Schumer (D-N.Y.) cited a report claiming that one recent virus came "within four hours of taking down the Internet" and that the four most recent viruses caused $12 billion in damage.

No one—including Clarke, the White House cyber security director—batted an eye to question the data. But if you took the figures at face value, these four *Über-viruses* supposedly caused fifteen times as much damage as the fully loaded Boeing 757 that crashed into the Pentagon (as noted earlier, total repairs cost $800 million and required a year to complete).[11]

Viruses, or any kind of program that is simply released into the Net, are ineffective weapons because the computers an adversary most wants to disrupt—military, infrastructure, financial—are run by the people who are most aware of the threat. Once they detect a virus attack, they know how to respond, and the attacker cannot change the configuration of a virus once it is released.

Virtually none of the alarmists warning about an electronic Pearl Harbor have sweated the details of actually planning an attack on someone else's information infrastructure. If they did, they would understand a basic principle: You can do a simple attack against a lot of computers. Or you can do a sophisticated attack against a few computers. But it is really, really hard to do a sophisticated attack against a lot of computers, especially an attack that would achieve a meaningful military objective.

The fact is, we know something about how hard it would be to pull off a nation-wrenching cyber attack because U.S. officials themselves have considered something like it. Recall the NATO campaign against Serbia in 1999, Operation Allied Force. The goal was to get the Serbs out of Kosovo. The experience, much of which has been made public, suggests that pulling off such a cyber strike is much more complicated than it looks. Here's what happened.

In 1999 the United States found itself in a war over Kosovo, a province in southwest Serbia. Kosovo is important for historical reasons to the Serbian ethnic Orthodox majority. However, most of the population of Kosovo are Muslims. From 1996 to 1998 Kosovar separatists conducted a low-level war against Serbian authorities.

The Serbs cracked down on the Kosovars, and in February 1999, NATO issued an ultimatum to the Serbs to remove their forces and allow Kosovo to vote on autonomy. The Serbian regime, headed by war crimes poster child Slobodan Milosevic, used a tactic that had become the method of choice during the Balkans wars of the 1990s: ethnic cleansing. Serb paramilitaries went door to door, murdering hundreds of Kosovars to show they meant business, and sending the rest down the road with just the few belongings they could pile into a truck or carry on their back. Eventually 800,000 Kosovars fled to the neighboring countries, Macedonia and Albania.

On March 24, NATO began its military operation to compel Serbia to capitulate. The campaign was remarkable because it was the first time that a country had tried to carry out a war almost entirely from the air; U.S. leaders were loath to send ground forces.

But there was also an information warfare component to Allied Force. "IO was part of the planning from the earliest stages," recalls James Steinberg, using the military's term for "information operations." Steinberg was Deputy National Security Adviser at the time. He became familiar with information warfare weapons during the mid-1990s. As director of the Policy Planning Staff at the State Department, he was State's representative to the panel that reviews covert operations. Now, as a top White House aide, he was ready to consider using them in war. "We used IO partly because of the nature of the operation, which was to coerce Milosevic to give up Kosovo. Partly it was because the tools were available for the first time." The objective of the information warfare campaign, says Steinberg, was to split the Serbian elite and make it turn on Milosevic, seen by U.S. leaders as the root of the Balkans problem. "We wanted to create confusion and distrust among those around Milosevic."[12]

In one operation that made it into the press, the United States tried to peel off Serbian industrialists—crony capitalists is probably a better description—who supported Milosevic. Just before a bombing strike, the Serb factory owners received e-mail, fax, and cell phone messages telling them the attack was coming. U.S. and British forces had hacked the Serb telecom net. The implied message to the Serbian elites: Get rid of Slobo now, and get your troops out of Kosovo. If you don't, NATO will destroy your moneymaking machine.[13]

Steinberg reportedly told officials, "I want to see the rapid economic death of Serbia." According to some stories, planners put together a map showing the interconnections of the Serbian elites. The map served as the template for targeting specific individuals.[14] Other reports said the CIA had a covert plan to hack bank accounts Serbian elites held in Switzerland, Cyprus, Greece, Russia, and China.[15] "We went through the drill of figuring out how we would do some of these cyber things if we were to do them," said one officer who was interviewed after the war. "But we never went ahead with any."[16]

. The Serbs did withdraw from Kosovo, and later, the Serbian elites turned on Milosevic and delivered him to an international tribunal at The Hague. But almost no one thinks this was the result of the information warfare campaign. Even Steinberg believes that the NATO success was more the result of the Kosovar resistance planning a ground war and Russia telling the Serbs that they would not assist.[17]

Why wasn't information warfare more effective? One problem was planning. Vice Admiral James Ellis, who commanded the U.S. air campaign in the war, said information operations were "perhaps the greatest failure of the war." Ellis said that the people responsible for planning information operations were too junior and did not have access to the top commanders.[18] So information warfare was never part of the top-level military planning.

Another problem was that the United States lacked the right tools for the job. When the White House asked for options, they encountered what Steinberg recalls as the "hammer and nail" problem—when you have a hammer, the whole world looks like a nail. While there were lots of cyber weapons for supporting military operations, no one had the responsibility for developing tools or plans to put suspicion and despair into the heads of Serbian leaders. So when the White House asked for ideas, the military came back eager to act, but with nothing that would meet their needs.

That was only natural because, remember, the Army, Navy, and Air Force build systems and train people to address *military* operational requirements. "When we hear talk of information warfare," General John Jumper later said, "the mind conjures up notions of taking some country's piece of sacred infrastructure." But, continued Jumper, that was "hardly relevant to the commander at the operational and tactical level."[19]

As a result, when U.S. leaders looked in the toolbox for the screwdriver to twist Milosevic's mind, all they found was that hammer. The Air Force, Army, and Navy had lots of ways to spoof or jam a radar network so a bomber could pass by safely. They even had psychological warfare materials designed to make a Serb airman or soldier want to go back home to his mommy in Novi Sad. But no one wearing a uniform was thinking about a plausible way to make the owner of a steel factory in Smederevo worry about losing his stash in Nicosia.

A larger problem was that it was never really clear how all the hacking and spoofing or even the less exotic operations were supposed to achieve anything—the basic question of *if we do x, then as a result y will happen.* Most of the operations that were on the shelf had been designed by the Army, Navy, and Air Force. No one had responsibility for designing information operations aimed at a strategic objective, like overthrowing a government.

The day after I spoke to Jim Steinberg about all this in Washington, I drove down to Langley Air Force Base—about an hour or so away, depending on traffic—to get the military's side of the story about Kosovo from officers at Air Combat Command. ACC is the parent organization to the Air Intelligence Agency and the Joint Information Operations Center, two organizations that have hands-on responsibility for developing tools for breaking into an opponent's electronic systems and fabricating the data to be inserted inside.

"We could use more guidance from the top," said Lieutenant General Bruce Wright, currently vice commander of ACC.[20] Wright reflected on the information war that accompanied Operation Allied Force and conceded that the Serbs often had the lead on the United States. Wright, a fighter pilot by background, goes by the handle "Orville."

When the Air Force wants to put money behind a program, it has to find a requirement it can match. But the military planners who draft the requirements don't talk to the White House.

"An Air Force O-5 isn't going to come up with a plan or a program to depose a national leader," one of the lieutenant colonels at Langley said, referring to his rank by its pay grade. There are four or five layers of bureaucracy separating the National Security Council from the working stiffs (their term, not mine) at Air Combat Command. So the Air Force never gets around to developing the plans and building the weapons that address the needs of officials like Steinberg. And when the White House asks for plans and weapons, there's nothing there, or at least nothing designed to meet their needs.

"There's no location where we can bring together all the different capabilities to shape the enemy's perception," according to Major General David Deptula. Deptula, who was one of the principal planners of the air campaign in Desert Storm, currently directs ACC's plans and programs. Deptula would have to sign off on any Air Force project for a cyber weapon designed to cause, as we now phrase it, "regime change."

In addition to all this, during the war in Kosovo the Pentagon was not even certain what kinds of information warfare were legal. This might not seem like a big deal, but it was. It goes back to an incident in the Vietnam War involving General John D. Lavelle.

In 1971 Lavelle was the commander of the Seventh Air Force and was directing the U.S. bombing campaign in Vietnam.[21] A fighter pilot by training, Lavelle fit the classic mold and believed in the classic theory of warfare and warriors: Win the battle, protect your people, and don't complain—just get the job done.

Alas, politicians had a more nuanced theory of war. Lyndon Johnson and Richard Nixon both tried to calibrate military operations with a micrometer. The politicians were trying to extricate themselves from a war they believed was unwinnable, and didn't want some air strike screwing up their carefully calibrated negotiations at a delicate moment.

The issue was the rules of engagement—when a U.S. warplane could attack a North Vietnamese air defense site. The rules said that you could not fire on a missile site unless it fired on you. At the same time, commanders running the war were pressing Lavelle for results. They wanted him to attack more aggressively. In other words, the President and the uniformed military had goals—and rules, and instructions—that were mutually incompatible.

As it happened, the specifics of the Lavelle case hinted at cyberwar issues that would emerge almost three decades later. Lavelle's problem was that the North Vietnamese began to link their air defense sites into a network. One radar or missile site could tip off another when an American aircraft was heading in its direction.

So Lavelle stretched his rules of engagement to the limit. He not only authorized his pilots to return fire when a missile site fired on them; he also told them to take out any air defense site that seemed to be assisting. Since any node can, in principle, support any other node in a network, Lavelle was essentially declaring open season on North Vietnamese air defenses.

Down the chain of command, intelligence analysts figured out the game and started to look for any argument they could muster that justified a target under the rules of engagement, as interpreted by Lavelle. Word got out, and a sergeant wrote to his Senator claiming that Air Force personnel had been told to falsify target descriptions in their reports.

Attacking unauthorized targets to protect his pilots was merely a venial sin. But allowing his men to fudge their reports was a mortal sin. Lavelle was relieved of command and demoted.

The Lavelle case is a classic in military ethics courses. The lesson: Lavelle was admittedly in a tough spot. Nevertheless, he made the wrong choice. Instead of trying to slip deftly through the gauntlet of conflicting orders and conflicting demands, he should have refused to fly until his bosses made clear exactly what they wanted him to do—and took responsibility for the risks. Ever since, the message has been clear: Orders must be obeyed. Lying is not acceptable, nor even fudging the truth. On the other hand, every soldier and officer has the right to demand that orders be crystal clear.

This is why, as we saw in the story of the Predator trying to track Mullah Omar, today we have a new component in our command and control systems: the lawyer in the loop. During the air war over Kosovo, sometimes a pilot would be flying over his target, ready to send his smart bomb on its way, only to wait until the targeteers and JAG could decide whether they were violating a policy directive or some article of the Laws and Customs of War.

No officer was going to engage in cyber war in Serbia until the civilians decided whether computers were fair game and which were off limits, and when American officials decided to attack Serb computers, they hadn't finished the legal work. Even before the Kosovo war, military planners were complaining about the fuzziness and contradictions. "I can bomb and blow it to smithereens," said one officer, referring to an enemy communications center. "I can bomb it, but they don't want me to manipulate the ones and zeros."[22]

So the Office of General Counsel studied the problem, putting together a team of lawyers from the general counsels of the Army, Navy, Air Force, the National Security Agency, and the Defense Information Systems Agency, the JAGs from each of the military services, and the legal counsel to the Chairman of the Joint Chiefs of Staff. Their conclusion: Information warfare is not that different from warfare in general, and as long as you avoid third rails like attacking under a flag of truce or misrepresenting your nationality, you can use ones and zeros as armed self-defense. The report said, "There are no 'show-stoppers' in international law for information operations as now contemplated in the Department of Defense." Alas, the report was finished in May 1999, just as the war was winding down.[23]

The Defense Department drew many lessons from Kosovo and improved its own planning. But it is still unclear whether the United States is prepared to use computer warfare to achieve political objectives. One lesson of the war in Kosovo is that the White House needs someone with responsibility for tying together all the pieces of an information war at the national level. He or she would coordinate the official U.S. message to the world from the public affairs offices in the executive branch, overt propaganda like Voice of America, covert propaganda—and plans to attack enemy information systems at the strategic and tactical levels. This White House official should also be responsible for addressing the legal questions surrounding these issues.

But return to the whole idea of electronic Pearl Harbor—a full-bore attack on the United States with catastrophic political, military, and economic results.

Many people speculate about how they can cause electronic mayhem. Few offer a logical explanation about how the mayhem will achieve a useful effect. Suppose Hezbollah crashes the power grid in California for a day, does anyone think we are going to change our policy about who owns the West Bank? Rethink our policy on Iran?

The scare mavens also overlook the obvious when they use the Pearl Harbor metaphor: *Japan lost World War II.* Thanks to Pearl Harbor, the United States had to kill hundreds of thousands of Japanese soldiers and sailors. We firebombed and nuked Japan's cities and overthrew its government. We made the emperor admit to being mortal and hanged Tojo. And in the process, thousands of Japanese were left broke, homeless, and reduced to eating dogs to avoid starvation.

No country would want to replicate Pearl Harbor—electronic or otherwise—because no country wants to replicate what happened to Japan. And that's why electronic Pearl Harbor makes no sense. It is hard to imagine how any of the scare scenarios describing a bolt-out-of-the-blue cyber strike would help a country, terrorist organization, or anyone else achieve a strategic objective. Sure, you can take out a computer running a power plant—you can also torpedo a ship, bomb a tank, shoot a soldier, or slit the throat of an unsuspecting sentry. But an effective weapon and a vulnerable target set do not equal a strategy.

Information warfare—in particular, attacking your enemy's computers—makes sense only as part of a larger plan. You have to be able to show how it will help you win a war, or put your opponent at such a disadvantage in the race to the end of the OODA loop that he simply gives in. Just as important, you have to show that attacking your opponent's computers is better than the alternatives.

Chapter 14

LES GUERRES IMAGINAIRES

So, if electronic Pearl Harbor seems unlikely on closer inspection, what would a real information warfare attack look like?

Defense experts have been writing about hypothetical wars of the future for centuries. It's a common literary device known as *les guerres imaginaires,* or wars of the imagination. The biblical tale of Armageddon is an example. So is H. G. Wells's *The War of the Worlds.*

More recently *les guerres imaginaires* have been the tool of choice for pundits and retired military officers who want to warn the public about some foreign threat they feel is getting insufficient attention. Hector Bywater, a British journalist who wrote for the *Baltimore Sun* and *The New York Times,* gained some fame after the raid on Pearl Harbor because his 1925 novel, *The Great Pacific War,* supposedly predicted the Japanese sneak attack. In the late 1970s, John Hackett, a retired British general, borrowed some scripts from NATO exercises to describe a 1985 Soviet invasion of Europe in *The Third World War.*[1]

The typical story line of a future cyberwar has airliners colliding in midair when air-traffic controls get hacked, financial chaos because banking computers go awry, and, of course, the *de rigueur* shutting down of power grids—the usual stuff that Roger Molander and Peter Wilson were writing about almost ten years ago. Even the cyber-security plan that the Bush administration released in September 2002 included sidebars depicting such scenarios.[2]

What is interesting is that most of these stories are written from the point of view of Americans being attacked. It's more dramatic that way. But what if we were to turn the tables, and put ourselves in the position of an adversary? How hard would it be to attack the United States? More important, what kinds of attacks would make sense?

Let's assume the Chinese decide to invade Taiwan, and the United States must decide whether to stop them. Professional military journals from China often discuss computer network attacks as a military option. Usually they describe U.S. capabilities and plans, but any military organization that has discussed foreign plans for information warfare has probably considered the option itself.

Chinese officials would likely design the attack to unfold on several levels. The first level would probably look something like "Hack the U.S.A. Week," a cyber rumble that occurred in March 2001 when Chinese Air Force Lieutenant Wang Wei flew his fighter jet into an American reconnaissance aircraft. The Chinese government announced that it planned to keep the aircraft and its crew indefinitely, and the Chinese media went wild protesting the supposed American aggression into their airspace.

Some Chinese technogeeks decided to make their contribution to the anti-imperialist effort, too. They announced plans to hack into as many U.S. computer systems as they could during the week between May Day and the anniversary of the accidental U.S. 1999 bombing of the Chinese embassy in Belgrade. No one really seemed in charge. It was mainly a grass-roots movement that the Chinese government—which ordinarily monitors use of the Chinese Internet closely—simply did not bother to stop.

Soon the Western media picked up the story. Inevitably, *Wired,* which covers just about everything concerning life on the Internet, was first, announcing that Chinese hackers planned a "week-long all-out crack attack on American Web sites and networks." The FBI issued an alert, and its National Infrastructure Protection Center (which monitors computer crime) reported that American systems operators had detected an uptick in "probes"—attempts to access their networks that seemed like something other than normal communications.

The FBI also reported that a new virus, Tiger, had appeared in some U.S. networks. The Tiger virus set up DDOS attacks and stole passwords from infected computers, mailing them back to China. And a few Web sites got painted with graffiti. This happens when a hacker manages to sign on to a network and then figures out how to fool the network into thinking he is a legitimate user of the system. Ultimately the hacker tries to make the network think he is the system administrator—this is called getting "root access." Then he can make the network do whatever he wants.

This kind of hacking—probing from outside and figuring out how to get in—is as much art as science. Hackers have lots of different tools and techniques. The basic principle is to use a feature that makes a computer perform a legitimate function, like accepting an e-mail message, and use it to make the computer carry out your own illegitimate function, like uploading a "Trojan" (a program that gives a hacker remote control of the computer).[3]

Naturally, software is designed to prevent this kind of hijacking. But software is complex, and it is impossible to understand every possible combination of commands and logic. So almost all software has flaws—so-called holes—that allow intruders to submit these hostile commands. Word gets around—hackers have their own online community—and over the years a compendium of tricks and holes has accumulated. Some holes become useless as software vendors develop patches, and others are replaced as the hacker community learns about new vulnerabilities.

With these techniques and know-how, hacking is a step-by-step exercise, somewhat akin to rock climbing, where you find a ledge, outcropping, or crevice to jam a pin into (or your foot, fingers, fist, or whatever else works). Then you swing your body to the next position and repeat the process until you get to the top. It's like solving a puzzle. You know the general contours of the wall, and adapt your strategy as you learn the details of the situation.

Similarly, a hacker knows the general contours of a computer network. Most use some version of the UNIX operating system (or a close cousin, like Linux) or a Microsoft product like Back Office. They know where the system needs to connect to the outside world to perform its functions. After that, it's just a matter of deducing the details as you move along, and then figuring out how to use each step to take you to the next one. If the hacker gets root access, to prove his success, he pulls the file defining the Web site's home page. He then adds his own words (usually scatological) as a sort of "Kilroy was here." During Hack the U.S.A. Week, the Chinese hackers sometimes added a picture of Lieutenant Wang.

It's impossible to stop these kinds of shenanigans without walling off every computer and electronic device from the outside world, which, of course, would make them useless. Also, many of the tools hackers use are widely available because they have legitimate uses. For example, automatic dialers were developed for telemarketers (admittedly, a somewhat dubious "legitimate" use), but a hacker can use one to search for a telephone line connected to modems. Alec Muffett, a programmer from Hampshire, England, wrote a program called Crack so the operators of computer network systems—"sysops"— could make sure everyone on their network was using strong passwords, but hackers use Crack to, naturally, crack weak passwords.[4]

There are many other examples. For instance, once installed on a PC, commercially available programs like Timbuktu allow an outsider to run a computer by remote control. They are a handy tool for a sysop who works in the company's main office in Detroit but needs to maintain computers in a satellite office in Skokie. But the technology behind the software is not much different in function from a Trojan (Back Orifice, one of the most famous Trojans, performs exactly this function, but covertly).

The list goes on. Back in 1995, Dan Farmer designed a program called SATAN. Farmer originally intended it as a security tool that allowed a sysop to check an entire network and detect any settings that had not been properly configured to keep out intruders. But a hacker can use the same tool from the outside to identify all the weaknesses in a system.[5]

These are all examples of a general principle. Software is like a big extended family tracing its toolmaking roots back to *Homo habilis* and the Olduvai Gorge. Most programs are derived from some other program. Often the connections seem unlikely, but they are there.

For example, word processors and Internet browsers share a common gene. Left to itself, your computer screen would display just lines of plain, unadorned electronic letters, like the letters in the IBM logo, which was, in fact, based on how computer text looked at one time.

Then, around 1983, Steve Jobs and Apple borrowed an idea from the Xerox labs—instead of plain letters, a screen would display lettering that looks like it came off a printed page. The software running the computer was designed so that, for each letter of text, it called up a small font file and inserted the file in place of the letter.

The name of this technology borrowed a punch line from an old routine the comedian Flip Wilson used to perform, and is called WYSIWYG (What You See Is What You Get). Apple introduced it to the mass market with its Lisa computer, an offshoot from the tree that eventually led to the Macintosh. Other software developers adopted the same trick, so instead of text on your screen looking like the IBM logo, it now looks like Arial, or Times New Roman, or whatever.

But WYSIWYG can be used for more than just easier word processing, because this "tagged text" technology can call up a file from a computer's own hard drive or from the drive of a computer thousands of miles away. This is the basic principle Tim Berners-Lee used to invent HyperText Markup Language (HTML) in 1989 when he was working at the Center for Particle Research in Switzerland. Click on a letter, call up a file—that is how Internet browsers like Netscape Communicator and Microsoft Internet Explorer work, and it is the basis of the World Wide Web.[6]

This is why it is impossible to eliminate or control hacking or, for that matter, computer warfare, without curbing all information technology. Software is like firearms, nuclear power, or airliners. They can all be used for legitimate or nefarious purposes, and it is hard to make the technology available to the many who would do good without making it available to the few who would do evil.

When the press announced Hack the U.S.A. Week, some of our own American technogeeks decided that they would return the favor and hack some Chinese Web sites. But none of this amounted to a hill of beans. It was a food fight, not a geopolitical death match. The sites that "got owned" during Hack the U.S.A. Week were just Web sites that various organizations use to provide information to the public. As it happened, the Americans came out ahead; defaced Chinese sites outnumbered U.S. sites by a hefty margin.

Hackers who paint graffiti on Web sites get a lot of attention, probably because you can show the results on television or in a magazine.

(Back in September 1996 someone—apparently Swedish, but you never know—hacked an unclassified CIA Web site and rearranged the letters to spell "Welcome to the Central Stupidity Agency.") But such incidents are just an embarrassment to sysops, who are exposed for having sloppy security.

The greater danger is from the hackers who don't want attention. And that would include the people behind the second level of the Chinese cyber attack. This second level in a cyberwar would be more serious—a professional computer network attack. The Chinese would have created a team of sysops devoted to this purpose. It would probably be a secret agency unknown to the public, and would involve maybe fifty to a hundred experienced network operators and software developers.

A minor part of the team's attack plan might consist of sophisticated viruses. The difference between these and the common viruses one finds every day on the Internet would be that the Chinese would design these viruses from scratch, and probably spend more time testing their behavior than your garden-variety hacker could. Their goal would be to develop code that most virus protection software couldn't recognize. Also, the Chinese would tailor the code so that the virus gravitated toward systems that were less likely to be protected, but still essential to the U.S. war effort.

For instance, it would be hard for the Chinese to get a virus into the secure networks that only the military and intelligence communities use. There are too many air gaps. Even if they did get the code onto a secure computer, unless they had a mole, they would not know enough about the network to make the virus propagate effectively.

But the Chinese might be able to infect enough commercial computers on the Internet to jam up a few companies that were important to the war effort—for example, a business-to-business Web site that kept inventories of aircraft parts. The Chinese would also try to compromise unclassified government Internet servers. Depending on how well the virus worked, it would basically keep government workers from using e-mail, and prevent officials from putting information up on Web sites to keep the public informed.

But as you begin to spin these virus scenarios, you soon begin to appreciate the problem facing the Chinese. As we have seen, to have a significant effect, you need a more and more complex virus, but complexity makes the attack less reliable and easier to detect. That's why most of the "professional" attacks would not be viruses, but highly focused efforts aimed at specific computer systems.

Working over several months, the team would identify computer systems in the United States critical to a military operation to assist Taiwan—for example, automated inventory databases used by shippers at ports like Long Beach and San Francisco. The computer attack team would then lay the groundwork by probing them quietly over time.

First, the team would "footprint" the networks they targeted for attack. This would include building a systematic blueprint of its design and vulnerable points. The team would have a catalogue of identified holes that exist in popular software packages, and they would know the slip-ups network operators commonly make in maintaining firewalls and other security measures. They will also know from their own experience the shortcuts taken by sloppy or lazy operators. The team would also include people familiar with odd or obsolescent operating systems, and at least a few specialists good at figuring out how homegrown programs work.

Once the team penetrated a network, they would lay trapdoors—hidden programs or software holes that would make it easier to get in next time. But their ultimate goal would be to have a complete set of mirror-image technical manuals that described how the targeted network operated, and a documented collection of procedures for cracking into the system and taking it over.

To disguise their probing, the team could use a common virus attack like Code Red or Nimda as cover. They could either launch one themselves, or time their probes to hide among the routine shenanigans of the international hacker community. Their probes would be just one more ping among millions of scans.

"That's exactly the way I would do it," says Duane Andrews, now a corporate vice president for Science Applications International Corporation, a major provider of computer security services. He even thinks that this may explain the huge volume of hacking that routinely occurs over the Internet. "You have a lot of activity out there," Andrews continues. He doubts that it is just teenagers with computer skills and too much time on their hands. "If it's just pranksters," he says, "then who's paying for the pizza and Cokes?"

Even so, serious computer network attacks involve methodical work, a lot of planning and organization, and a quality staff. And the most important parts of the computer network attack team would probably be the traditional intelligence service that supports it, and this should tell you something about cyber threats.

It is usually hard to crack a properly run computer network from the outside—at least, not without getting caught. Setting up the attack requires too much probing. And the computer systems that one most wants to crack usually have the best security. So if you are serious about computer warfare and plan to rely on it, you need to develop insider access.

Besides, computer warfare is rarely an all-or-nothing situation. It's a game played in shades of gray, where success on both the offense and the defense depends greatly on details. A few years ago, *eWEEK*, a trade newspaper, illustrated this fact when it started running a competition it called Openhack. (The name was a play on "open source software," programs like Linux and Apache that are in the public domain and maintained by volunteers.) The newspaper set up a computer network designed to be tough to break into, and then invited all comers to try their skills.

By the third round, about 200,000 people reportedly made at least some effort to break into the system, registering 5.25 million hits in the process (it was hard to tell what constituted a serious attack; these were the raw numbers). In the first two rounds, a hacker was able to download and alter the database on the server—the ultimate test of whether a system is not secure.

In the third round, though, *eWEEK* used a "trusted operating system," or one that is heavily compartmented and requires several people, each with unique operating privileges, to run. This approach was developed in the early 1980s by some private organizations with advice from the National Security Agency.[7] The Openhack trusted system emerged unscathed. Some hackers did manage to get into one part of the system or another, but no one was able to compromise its operation.

The only problem is that trusted systems are more complicated, and more costly, to operate. And that is the real challenge in defeating serious, determined attacks. Security is almost always attainable, but only at a cost. Protective measures run the risk of making a system too inconvenient or too expensive to use. It is a straight trade-off. Think of it this way:

Imagine two worlds. In one world, all electronic devices are computers that, except for their identification numbers and Internet addresses, are identical. Chips, storage media, peripherals are all the same. All run the same software, all are installed with identical settings. What is more, all of the computers are hardwired so they are always connected to a worldwide network, and the address of every computer can be looked up in a directory that identifies its owner and all users. And every computer has the same password: PASSWORD.

The great thing about this world is that it is as user-friendly as one could imagine. Once you learn how to use the standard computer, you can use any electronic device that happens to be handy. No more need to read instruction manuals. No more training. Also, everyone can easily communicate with everyone else. Exchanging data is easy, and no one loses any time fumbling for passwords or trying to make two systems interface.

The main drawback with this world is that there is zero security. Anyone can access the data on anyone else's computer, anywhere in the world. Also, this world has zero resistance to viruses. Anyone can write a simple program that propagates instantly around the globe and crashes all computers at once. This is a hacker's dreamland.

Now imagine another world, the opposite of the first. All electronic devices are unique. All the chips, storage media, and peripherals are tailor-made for each machine. Every device has a one-of-a-kind software suite coded by a programmer from Peshawar that no one else has heard of. No device is connected to any other device, and no computer has a floppy disk, tape reader, CD player, or any other kind of removable media. And every device has just one authorized user, who must memorize a 56-digit case-sensitive password consisting of numbers, letters, and symbols that must be entered when logging on and after three minutes of inactivity.

This is the world that security officers dream about. It is totally, absolutely secure. The only person who can operate a particular machine is the person authorized to use it. It is impossible to take over anyone else's computer or make it malfunction because, even if you knew how to work it (which you don't), it lacks removable media or an electronic connection, so you cannot get any data into it—or out of it.

The only problem is that all the computers in this world are useless. No one can run anyone else's data. Every time you want to move to a new job, you have to learn how to run a computer from scratch. Also, because every computer is a unique, handcrafted work of art, it is incredibly expensive.

The lesson: The perfectly usable computer has zero security. The perfectly secure computer has zero usability. In the real world you have to find a compromise between the two. It is precisely the features that make computers and all other information devices versatile, easy to use, and efficient that also make them easier to attack. This is why computer security—especially against serious, well-funded, sophisticated attackers—is such a challenge. But the fact that there is so much variety and complexity in the cyber world is why pulling off an effective attack is so challenging, and why insider knowledge is so important.

The reason, incidentally, why the process of attacking a computer network to take it down is so well understood is because it is the same process an attacker would use to break into a computer system to commit espionage. The main difference is strictly whether you want to lurk and listen or crash and burn. Once again, there's the Perennial Question for information warriors: *Deny, deceive, destroy, or exploit?*

In any case, in future wars attacking computer networks will likely become a routine part of military operations, mainly because it's an obvious tactic and there is little reason not to. Many foreign military leaders have already acknowledged that they are exploring the option, and the military literature in China, Russia, and India—to name just three countries—discusses it frequently.

In a real war, the effectiveness of a computer network attack would depend greatly on how much effort the attacker was able to devote to the operation. High-quality hacking is labor intensive. There would be some hits and misses. So a telephone system might go out here and a power plant there. The successful operations would most likely resemble some of the worst cases of computer vandalism or glitches we've seen in the past, like the May 1998 failure of PanAmSat's Galaxy 4 communications satellite. The satellite's computer crashed unexpectedly, and the spacecraft temporarily went out of control. Somewhere between 80 and 90 percent of America's 45 million pagers went dead, and National Public Radio lost its feed to local stations.[8]

Yet none of the activities described up to now had anything to do with the main Chinese information warfare attack. They were just incidental events—either a sideshow, or a neat trick some general wanted to try out. Let's get back to *les guerres imaginaires*.

The main computer network attack took a year to plan, two years to implement, and about thirty minutes to execute. The most ironic thing was that even many years later, no American knew it had occurred. The U.S. government would spend months investigating the kinds of cyber attacks described above. A blue-ribbon commission would issue a 200-page report analyzing their impact and proposing preventive measures for the future. But the report never mentioned the real attack. Here's what happened.

Chinese officials had learned through experience how a confrontation over Taiwan might play out. They especially wanted to avoid a replay of the U.S.–Chinese clash of 1995–96, when Taiwan's president Lee Teng-hui visited the United States. Lee had been invited to make a speech at his alma mater, Cornell University, in June 1995. Chinese leaders believed the visit looked too much like official recognition of the government on Taiwan.

The Chinese government expressed its displeasure with a series of military exercises designed to intimidate the Taiwanese. In one exercise, the Chinese fired six ballistic missiles that landed just eighty miles off the coast of Taiwan. The Taiwan stock market promptly fell in value by almost 7 percent. As Taiwan's presidential election came around the following spring, the Chinese tried to increase the pressure on Lee with more missile tests and landing exercises on nearby islands.

While all this was going on, the United States tried to convince China that it would defend Taiwan. Two American aircraft carriers moved inside the waters separating the island from the mainland. Eventually tempers cooled down. Lee made clear that he did not intend to declare independence for Taiwan. U.S. leaders made clear that they wanted good relations with China. China ramped down its military exercises.

But Chinese leaders had learned a lesson from the episode: If they intended to settle the Taiwan issue with a military attack, they had to have a way to put the United States off balance, at least enough to give U.S. leaders second thoughts about aiding Taiwan. Chinese military planners decided they needed to launch a sudden, comprehensive missile strike against Taiwan's ports, airfields, and military command centers. This would make it much riskier for the United States to assist the Taiwanese.

The question was how to prepare for such a strike without alerting the United States. The Chinese needed a brief window to roll out and launch their missiles. According to their analyses, Chinese military planners believed that if they could just interrupt U.S. reconnaissance satellite coverage for thirty minutes, they would have that window.

The Chinese had analyzed U.S. satellite coverage of the Taiwan Strait using a commercial software program—say, Satellite Toolkit.[9] They concluded that there was a brief period when the ability of the United States to detect Chinese preparations for a missile strike depended almost entirely on a single surveillance satellite.

That satellite transmitted its data through a series of communications links, one of which happened to be a commercial relay satellite circling the globe in geosynchronous orbit over the Western Hemisphere. (Recall that most Defense Department communications, even highly classified ones, travel through commercial links.) The Chinese computer network attack team proposed crippling that satellite. It was sort of a man-made version of the PanAmSat failure.

Before Chinese political and military leaders would agree to approve the plan, though, the computer attack team had to convince them that they had met five criteria. First, they had to show that the plan was legal under the rules of war. Surprising as it may seem, most countries—even authoritarian regimes like China—are careful not to break the conventions governing the conduct of armed conflict. The main exceptions are a few rogue states. At a minimum, officials worldwide want enough obiter dicta from the legal beagles to cover their hindquarters if their operations go sour and they wind up in the dock. Lawyers are in the loop everywhere.

Second, the computer attack team had to prove that it would not cause excessive collateral damage. One reason electronic Pearl Harbor scenarios are so implausible is because such an all-encompassing, chaos-producing infotech meltdown would harm millions of civilians. Chinese leaders worried that indiscriminate havoc would be a violation of the rules of war. But, more important, the Chinese leaders knew their own military forces were heavy users of commercial satellite communications. The team had to make sure that no Chinese forces would be blinded by the strike.

Third, the Chinese computer attack team had to convince Chinese leaders that the attack would work—that it would perform as claimed, and the relay satellite would be neutralized. Chinese military officials would not let an important operation—in the case of retaking Taiwan, the operation of a lifetime—hang on a hit-or-miss scheme. Proving the attack would be effective was harder than it first seemed, because communications satellites are complex, and it is not easy to prove that you can shut down a particular satellite without actually trying it.

Fourth, the Chinese computer attack team had to convince their superiors that it was worth putting their penetration of a U.S. computer system—the one controlling the satellite—at risk. Remember that the process of attacking a computer network is usually the same as penetrating it for exploitation. The computer attack team had to show that the Americans would not be tipped off and fix the holes they had used for the attack, or, if they did, that the gains were worth the loss of an intelligence source.

And fifth, the Chinese computer attack team had to show that a cyber attack was the best way to carry out the operation. This was where the team stood on firm ground. It is simple to put a satellite out of commission—a satchel charge under the right uplink dish will often do the trick. But it is hard to disable a satellite in a way that is deniable, undetectable, and reversible. If those are the criteria, a cyber attack is often the best option.

The Chinese had already stolen the satellite's operating system by hacking into one of the labs that had designed it, just as someone had grabbed a copy of the navy's OS/COMET satellite software by penetrating a Naval Research Laboratory server in December 2000. The Chinese had also managed to get an agent inside the satellite ground station. The agent copied the appropriate uplink software and authentication codes.

At the critical moment, another agent operating from a satellite control station concealed on a ship anchored off the coast of Ecuador hacked the satellite and put it into a "safe hold." (Satellites are designed so that when a malfunction occurs, they shut down and point their solar panels to the sun and antennas to earth, so that their batteries remained charged and staff at the ground control center can figure out what went wrong and how to fix it.)

It took about three minutes for the U.S. surveillance satellite operators to realize and confirm they had lost their data stream. It took them another four minutes to contact the relay satellite operators to tell them they had a problem; by that time, of course, the operators were well aware of the emergency and were, to put it mildly, preoccupied. It took about fifteen minutes to reconfigure the surveillance satellite to communicate through an alternative link and test the connection. It took about ten minutes for the analysts using the satellite's information to sort through their data and pick up their analysis from where they left off. In other words, the Chinese had two minutes to spare.

That was enough time for the Chinese to wheel out their missile launchers from their shelters and fire off a volley that destroyed 100 key targets up and down Taiwan. In the ensuing chaos, the Chinese followed up with additional missile strikes and aircraft attacks. By the end of the day, they were ready to launch their cross-strait assault, occupy a beachhead, and present Taiwan, the United States, and the rest of the world with a *fait accompli*.

At that point, U.S. leaders had to decide whether they wanted to go head-to-head with China in a war on the other side of the world. It would have required U.S. forces to operate at the end of long supply lines and without a staging base on Taiwan itself. It was easier for everyone to negotiate a peace agreement that gave Taiwan roughly the same status and guarantees as Hong Kong.

An implausible scenario? Recall Task Force Normandy, the helicopters that destroyed those Iraqi radars at the beginning of Desert Storm. Central Command planners had originally considered three options: Using special operations teams to destroy the radars, using special operations teams to guide helicopters to the targets, and using fighter aircraft to bomb the targets. None of the options were wholly satisfactory, because they all had the potential of tipping off the Iraqis, especially if the target was not completely destroyed and survivors called back to Baghdad.

But if it had been possible to quietly hack into the Iraqi radar network, making it look like a normal malfunction, the American planners might have had a better option to create the hole they needed, just long enough to make the larger attack a success. That's what serious information warfare is really about. Information warfare is rarely an end in itself. It is always a means to get ahead of your opponent so that you can destroy him, or leave him so cornered that he will give up. Designing a computer network attack well requires a sense of nuance, and that's what the fear mavens who talk about an electronic Pearl Harbor seem to miss.

A Defense Department official who spends most of his time on information warfare planning—laying the groundwork for computer network attacks—said to me a few years ago, "Officials came to me saying 'I want to turn the lights out in Baghdad and send a message to Saddam.'" He rolls his eyes, like the colonel hosting us at NORAD.

"Sure, you could have taken out the power grid, and it would have been out for thirty-six hours—as though some of these countries weren't used to not having power for thirty-six hours! But the main message you would send Saddam is that we can take out your electricity, and then he would have fixed it so we couldn't do it when we really needed to."

But then he went on. "We've managed to convince ourselves that we are so vulnerable to computer network attack that we don't dare do anything ourselves. That's because we are always thinking about the big kill. There are subtle things you *can* do that make a big difference in a military operation."

Chapter 15

IF THERE WERE A FRONT IT WOULD BE HERE

When someone does decide to attack American computer networks, the beachhead could be companies like the one in a nondescript five-story white concrete office building on the edge of Alexandria, Virginia. Riptech, Inc., qualifies as being "inside the Beltway" by about fifty yards or so. Interstate 495 runs just outside its offices. In an era when office space is a commodity sold by the square foot, Riptech's home is indistinguishable from any of the other concrete boxes that dot the northern Virginia suburbs—what the state's promoters like to call the Silicon Commonwealth.

Riptech sells "managed security services." Companies and government agencies hire it to monitor their computer networks. Several other companies, like Genuity, Counterpane, TruSecure, and ISS offer similar services. But to understand how they work, you first need to know a little about how division of labor in the New Economy works.

When you log on to the Lands' End, United Airlines, or Citibank Web site, you rarely communicate with a computer inside the company's own buildings. Most large retailers and banks hire a company that specializes in Web site hosting. These companies operate "server farms"—large industrial computers on which the Web site you see on your own screen is hosted.

This is because server space, like office space, is a commodity. The product is pretty much the same, no matter whom you buy it from, so companies compete in service, reliability, and price. Each company in this chain—everyone from the realtor renting the office, to the Web site host, to the company renting the server—specializes in what it does best. The idea is to build it in bulk, offer a satisfactory product, sell it cheap.

Riptech found its own niche in this chain. It monitors networks by long distance from its single Alexandria office. It really doesn't matter if the cable connecting a monitor and keyboard to a computer is five feet long, five hundred yards long, or five thousand miles long, because the ones and zeros are zipping along at the speed of light. So a company like Riptech can watch more than one thousand computers or computerlike devices for more than four hundred customers spread across the country.

Riptech's technicians are linked to the computers they are monitoring via a software program that watches the firewall on each machine. This software, the key technology that Riptech offers, dates back to the founding of the company. Tim Belcher and Elad Yoran, two former Army officers, originally established Riptech in 1998 as a consulting firm specializing in "computer forensics." If a hacker got into someone's computer, Riptech searched the machine's logs to try to trace the violator and figure out how he did it.

Riptech eventually developed a program that automated the analysis; you put the software on your computer network, and if someone hacked you, the program spit out the diagnostics in an easy-to-analyze form. From that, it was just a short step to send the diagnostics back to Riptech over a data link the instant they were collected, and its computer monitoring service was born.

By 2002 Riptech had about 160 employees. When I visited the company, I met with Bruce Artman and Joe Pendry. Artman goes by "senior global security architect" and Pendry handles external relations, but just about everyone at the company has some kind of technical security background.

The heart of the company is its Security Operations Center (SOC), a room that looks something like NASA's Mission Control in Houston. It has several big-screen displays at the front and four rows of desks with computer terminals where analysts monitor their clients' networks. (When I was there, one of the big-screen displays was tuned to CNN, which seems standard viewing in most control centers I've seen lately, except when the boss is a Republican and the station is set to Fox News.)

You enter the SOC through ten-foot-tall bronze double doors after your escort punches in a code on a control pad mounted on the wall. Pass through the doors, and you are greeted by an identical set of bronze doors, which you can't go through. They lead to the main floor of the SOC, and only actual operators are allowed inside.

But if you look to the left, you see a mini-theater consisting of about twenty leather seats, much more plush than what you see at a typical government facility. You can sit and watch the operators in the SOC through a glass wall. It's a good show. With a little imagination, you can pretend you are not just at the SOC or even NASA's Mission Control, but on the Starship Enterprise, preparing to do battle with the Klingon Empire or the Borg or whomever it is that the Trekkies are worrying about these days.

Artman admits the SOC was styled partly to impress Riptech's customers. But the company is a serious operation and has some features that set it apart from most companies that offer similar services. Thanks to its software and customer base, Riptech can see an actual cross section of attacks against a variety of targeted computers.

Most reports on computer crime that you read, like the one from the Computer Security Institute that the FBI sponsors, are based on surveys. Somebody sends a questionnaire to a few hundred companies and asks them how many times someone has tried to hack into their network and how much it cost them.

One can be certain that it's not the CEOs of these companies that are filling out those forms. For that matter, there is no real way to know who is filling them out or how much time they are spending on them. And as for the damage estimates, we've already seen that they're suspect at face value.

Riptech takes millions of logs from thousands of firewalls and uses "data mining" to detect patterns and trends. Because Riptech's statistics are based on actual logs from actual computers, they are linked to real-world activity, not someone's impression or memory. This makes them interesting, if not necessarily comprehensive. Artman says they throw out virus attacks like Code Red, which just about everyone experiences and which would skew the results. They're mainly nuisances in any case. Riptech is more interested in serious, one-on-one attacks.

"The most interesting thing we discovered is an increase in targeted attacks," Artman says. "We found that 40 percent of attackers that showed up at one place were not showing up anywhere else." That is, Riptech is seeing a lot of cases where a specific hacker goes after a specific set of Internet addresses.

For example, Riptech might monitor a server farm that hosts corporate computer networks for Target, Wal-Mart, and J. C. Penney. The files and software for each might even be on the same disk drive. It is just as easy to attack sites with one set of IP addresses as another. But instead of attacking all of them, more attackers today seem to go after just a particular set.

Riptech's analysts can't track an attack back to a specific individual, because the attacker may be using a cutout—say, someone's computer that he has hacked in the past and is using now as a platform to hide his true identity. But Riptech can be reasonably confident that a series of hacking attempts come from a single source. They can recognize the originating computer because it will have its own address, or perhaps its own idiosyncratic pattern of attempts.

Put it together, and it means that Riptech is finding hackers who have spent the time and effort to look up the owners of a block of IP addresses, bird-dog them, and try to break into their Web sites. "Each attack has been reviewed by a human being," Artman continues, meaning that one of Riptech's analysts has verified the pattern.

The difference between this kind of precision attack and your run-of-the-mill virus is the difference between a sniper with an AR-50 putting several well-placed shots into a single target, and a gang armed with water pistols firing randomly into a crowd. Artman says that attackers are using the flaws in software packages that most hackers and security specialists know about. The only problem is, we don't know for certain the identity of these attackers, and in most cases we really don't know what they are trying to do. There are two reasons for this.

First, to deduce what a hacker is up to in an attack, it is necessary to go through the system logs, list every command he sent the computer, re-create the session, and then assess what his purpose was. Did he download passwords of customers who charged more than $1,000 a month—rich people, meaning that the hacker might have been a potential swindler? Did he steal credit card numbers of people who charged just a few items during the year, meaning that he was trying to harvest credit card numbers that he could use without anyone noticing? Or did he try to identify the buying habits of political figures—say, for counterintelligence or blackmail?

"There are precious few people doing analysis of the log data to see what's going on," Artman says. "It's this analysis that's the expensive part." Up to now no one has been interested in paying for this kind of analysis, and it isn't Riptech's business to do so either. It gets paid by companies to protect their computers, not play detective or intelligence analyst.[1]

Yet this is exactly the kind of analysis we need if we really expect to understand threats to U.S. computer systems. It is much harder than racking up sheer numbers of scans or asking survey respondents to make seat-of-their-pants guesses about the cost of these so-called attacks. But it would tell us what it is that we really need to fear.

The other reason why it is so hard to understand an attacker's goals, even if you review the computer logs, is that the attack could be disguised or stretched out. "An attacker can put in 'agents' who can report back intermittently," says Artman. He is referring to mini-programs that hide in a system by mimicking the appearance of regular files—data records, e-mails—but which really act like automatons to map out the system so that it can be attacked later. To avoid detection, the agents would transmit their information buried in the flow of data that the network normally sent to the outside world.

While part of Riptech is doing this kind of real-time monitoring, the other part of the company is doing the more traditional kinds of security consulting, like red-teaming their client's networks. These results are interesting too.

Pendry, who specializes in power systems, says, "We have found several times that we can get into their SCADA through their corporate site." Pendry is referring to the computer systems that monitor and operate factories, transportation systems, and utilities. (The acronym is pronounced "skay-duh" and stands for "supervisory control and data acquisition," but everyone knows that when you say "SCADA," you mean "industrial-strength computerized control system.")

"Every IT guy will tell you his SCADA is air gapped," says Artman, shaking his head. "Every time, we find some way of getting across." It's just too tempting to build in a link that lets you check in on your infrastructure from your office network. The designers try to hide the link, but, as Artman says, "Security through obscurity is no defense."

Even the best-intentioned system designer is apt to make a connection between SCADA and the outside world, if only by accident. For instance, if a system has a power generator, it probably has some link somewhere that measures how much fuel the generator is burning (that's SCADA) and downloads the data to the home office so that they can order some more (that's the corporate network). If there's a connection, and there is data passing through it, then it is at least theoretically possible to do the hack.

It's a lot harder than it looks, and it's infinitely easier with an insider to help. But, even so, Riptech has showen many a customer how Riptech technicians could figure out the path to the outside, jump the air gap, and fiddle around with the client's SCADA. "The customer didn't let us do it, but we showed them to their satisfaction that we could."

Suppose a hacker probing or manipulating the SCADA was a foreigner. Would Riptech notify the government?

"No way," says Artman. "We would take all the data, package it nicely together, and give it to our clients. But it's up to them to decide what to do."

Riptech's customers are often reluctant—understandably—to call the feds. Companies are wary about alarming customers whose electronic commerce may be less secure than one might hope. The first instinct of a company is to stop any damage and wait and see what is really going on. Alas, if the hacker is something more than a techno-geek—say, a foreign intelligence service or military organization—we have a gap in our warning network.

And there's the problem in defending the information infrastructure. If anyone is likely to spot the first signs of a serious cyber attack—and, as we have seen, that is itself a dubious proposition—it would likely be companies like Riptech. But they are not connected to the Defense Department or, for that matter, anyone else in the government. Add that to the fact that real computer attacks will likely be subtle, and the fact that companies have many incentives not to report hacking incidents and the fact that the government hasn't figured out its own reporting chain with clarity, and one begins to wonder whether a cyber equivalent of a radar defense parameter makes any sense at all. In any case, previous efforts have fallen short.

Government-sponsored organizations like Carnegie Mellon University's Computer Emergency Response Team Cordination Center (CERT/CC) are not designed as early warning, rapid-response attack-detection systems. They mainly deal with viruses and software holes that threaten the entire Internet. When the CERT/CC receives a report, it first tries to get a sample of the virus or faulty software to analyze the problem. It usually waits until it has a fix before issuing an advisory.

Similarly, the Clinton and George W. Bush administrations have both encouraged industries to establish information sharing and analysis centers (ISACs). But their function is mainly to allow companies to deal with a common threat (say, a virus keyed to the IT industry or a specific hacker after financial networks) by exchanging information in a private channel. ISACs are all financed by the companies they serve, not the government. So, like Riptech and other managed security services, their first instinct is to protect their customers, not spread word to a wider audience.

Add in the technical complexity of detecting a serious cyber attack and it boils down to this: Monitoring all the nation's computer and communications networks is incredibly complex and requires the active cooperation of too many parties with diverse interests. It's like controlling the nation's borders. Even after spending enormous amounts of money, there would still be many holes.

This is why our current approach to cyber security is so misguided. In September 2002 the government unveiled a draft policy for cyber security. It bore an uncanny resemblance to the draft plan unveiled three years earlier. The basic ingredients were the same: Federal centers to monitor network attacks and calls for more information sharing between the private sector and government authorities.[2]

Depending mainly on a government agency to coordinate a national cyber defense in real time is a fool's game. Computer operators need greater incentives to take reasonable measures to protect their own machines. Bill Crowell, who became the CEO of a Silicon Valley start-up when he left government, observes, "I never see security listed in anyone's 10-Qs of the companies I see," referring to the quarterly financial statements that publicly traded companies must file with the Securities and Exchange Commission. "That tells me the CEOs of those companies aren't thinking about security."

He has a point; the SEC requires companies to disclose what they consider to be "risk factors" that will affect their business. Companies apparently don't currently consider attacks on their networks to be a risk. As long as we are rethinking auditing rules in the aftermath of Enron, Worldcom, and Arthur Andersen, it might be a good time to think about adding cyber attacks. The government could, for instance, require an independent security audit for networks carrying financial records or intellectual property.

Ironically, this is the one approach that the government has avoided. For example, the word *liability* never appears in the new cyber security plan, and it mentions "regulation" mainly to assure everyone that the plan won't impose any new ones—even though it is hard to think of a mass-produced product released to the American public that is less reliable and problem-plagued than software.[3]

Read the label on your next software CD. It tells you that you are only buying a license to use the software on an "as is" basis, and that the company is not responsible for any loss of data, damages, injuries, and so on that the software might cause. Break the seal on the shrink-wrap, and you agree to the terms.

Imagine if the only way to purchase a Ford or Chevy was with a disclaimer saying that the car was sold "as is," carried no guarantee that it would work, and the manufacturer had no liability if it was defective! You begin to see the problem. Except that in this case many cars would even come with a nonremovable ignition key, so that teenage pranksters could take them for joyrides whenever they wanted. This is essentially the situation with software—and your computer, and your Internet service provider.

Securing the nets is a complex problem. There is no single solution, because software, hardware, and operators can all create vulnerabilities to attack and a sophisticated attacker will always look for the weakest link. And as we have seen, there is an inherent conflict between making a system usable and making it secure. Like all big changes undertaken by government, industry, and the public, the devil is in the details, and the solution is in incentives.

But companies themselves have proved they are willing to spend real money on computer security when they perceive that it is in their interest. Just as I was walking out the door at Riptech and the markets were closing, the company announced that it had been bought by Symantec for $145 million in cash. Not bad for a company trying to make it in the technology sector two years after the dot-com bubble burst and on the same week the Dow Jones Average hit a five-year low.

Chapter 16

COMMAND OF THE NETS

Security in the Information Age depends heavily on who has "command of the nets"—that is, who has greater control over the design, manufacture, and operation of information technology.

In some respects, "command of the nets" resembles traditional geopolitics, which is itself a fairly modern idea. The idea that a country could dominate the world by controlling a particular piece of real estate dates back only 200 years or so. Before that, geography was so rudimentary that it was hard to tell where anything was, let alone whether it was strategic. Most world maps had big blank spaces where no European had ventured. Even the parts that were filled in were iffy. With no geography, there can be no geopolitics.

This changed once the Industrial Revolution provided sextants, steamships, accurate timepieces, and reliable maps. For the first time, you could get to where you wanted to go, and know where you were once you got there. The second great era of exploration began.

So, in 1830, some members of the Raleigh Dining Club in London formed the Royal Geographic Society. In 1857 Queen Victoria granted the group a charter to pursue "the advancement of geographic science" and the "improvement and diffusion of geographic knowledge." By 1870 the society was well established and had enough money to buy a home for its headquarters at 1 Savile Row. (A few years later a group of Americans met at another dining club, the Cosmos Club, in Washington, D.C., to form their own exploration association; this became the National Geographic Society.)[1]

The Royal Geographic Society funded David Livingstone, Henry Morton Stanley, Robert Falcon Scott, Ernest Shackleton, and others who didn't mind polar cold, tropical rain forests, or tall mountains while trying to see how far they could carry a Union Jack from Britain. Meanwhile, back home, the society also paid academics to make sense of all the new information the explorers were collecting.

One of them was Halford John Mackinder, the eldest son of a middle-class doctor in the Midlands. Mackinder was a bright kid who was good at physics and chemistry, but his heart was really in the liberal sciences. With the support of the Royal Geographic Society, he wrote "The Geographical Pivot of History," an essay in which he tried to make the case that, at least as far as nations were concerned, geography was destiny.[2] Mackinder's argument was basically the same as the one real estate brokers use nowadays: What matters is location, location, location.

By Mackinder's reckoning, the most prime real estate of all was in Eastern Europe, supposedly the world's center of gravity. Whoever controlled Eastern Europe had the best access to materials, could move armies most efficiently. In short, Eastern Europe was the ultimate high ground, or as Mackinder summarized it:

> *Who rules East Europe commands the Heartland*
> *Who rules the Heartland commands the World-Island*
> *Who rules the World-Island commands the world*

Railroads were the key to Mackinder's theory. In Mackinder's time, the main way—often the only way—to get from point A to point B on land was to catch a train. Trucks and automobiles were just starting to appear, and it would be another thirty or forty years before any country had a modern highway system. In the 1890s railroads captured the public's imagination the same way the dot-coms would in the 1990s, and offered similar amounts of speculation, hype, swindles, boom and bust.

If railroads held a country together, then it logically followed that the country that pulled together the most land under a single integrated railroad system would have an economic edge. With a good railroad system, a country could move goods from factory to factory, and deliver finished products from factory to market. With a larger landmass, a country would have better access to coal, iron, and other raw materials, which also depended on railroads for transport.

Moreover, in the nineteenth century railroads were also how armies delivered their troops to the front. The American Civil War was the first railroad war and set the pattern. The Federal army was based in Washington and attacked the South from the railroad yards in Alexandria, across the Potomac.

The Confederates tried to use Richmond the same way, and the Federals' campaign to break the South was largely a campaign to capture its railroads. That's why so many Civil War battles have names like Brandy Station, Manassas Junction, and so on. As the Federals cut through the South, the Union troops would rip up the tracks, pile the ties into a pyre, and heat the rails so they could bend them around a nearby tree. (William Tecumseh Sherman's troops seem to have been especially fond of the practice, and during his March to the Sea the results became known as Sherman's Hairpins.)

All of the European powers had railroad-based plans for mobilization well into the twentieth century. Railroads were so critical to mobilization that armies usually took them over during wartime. Even as late as the 1980s, a Soviet attack on NATO would have depended on four rail lines that stretched across Poland. (One of the more esoteric arguments among military specialists was how quickly the Soviet army could transfer cars from the narrow-gauge Soviet rail system to the wider-gauge system that began in Eastern Europe.)

In effect, Mackinder saw the Eurasian landmass as one big roundhouse, optimally positioned to collect troops from the provinces and deliver them to the front. Of course, no serious analyst today believes that any power can dominate the world by controlling some real estate in the vicinity of Cracow. But Mackinder's thinking still permeates our ideas about strategic terrain. One of the biggest problems in U.S. defense planning is the vast distances U.S. forces would need to travel to reach the Persian Gulf and the Pacific Rim, the places we think our forces may need to fight. Mackinder would likely laugh, and dismiss the United States as a mere Rimland power struggling to protect its interests against Heartland powers like China, Iran, and Iraq.

Meanwhile, at about the same time Mackinder was extolling the importance of land—specifically, the Heartland—another pundit was pushing the importance of the other 78 percent of the earth's surface, the oceans. Alfred Thayer Mahan had graduated second in his class at the Naval Academy, but was an odd fit in the Navy and an unlikely candidate for patron saint of sea power. He loved sailing at a time when navies were converting to steam, and was cold and aloof in a service that rewarded the gregarious and personable. After an unremarkable career in the Civil War, he was directed to establish the Naval War College in 1873. Mahan did not know that he was, in fact, the Navy's third choice for the position.[3]

Even so, Mahan found his niche as a combination historian, popular writer, and serving naval officer. After reporting for duty at Newport, Mahan—who had not published much of significance before— began writing lectures and articles and soon became one of the best-known historians of the era. The British, and especially the Royal Navy, liked his work, which was only natural, considering that his best-known book, *The Influence of Sea Power Upon History,* focused largely on how British sea power had shaped events for 200 years.[4] His message: When the Royal Navy was strong, Britain prevailed, and when it was weak, Britain failed.[5]

The lesson: Britain—and by extension, the United States—needed a strong navy. Mahan coined the term "sea power" as a shorthand for having enough warships to keep the sea lanes clear for friendly commerce. Mahan claimed that sea power trumped land power, because nations and armies could not survive without access to the sea and global trade.

Mahan has not fared well among modern theorists; most think that his ideas are as overly simplistic as Mackinder's. But, as with Mackinder, many of Mahan's ideas are in the background when politicians argue over the defense budget. Navy officials themselves use Mahanian language when they publish doctrine about controlling sea lanes and the advantages of aircraft carriers, which let the United States send military forces overseas without having to depend on the cooperation of allies to provide bases.

The argument about whether land power or sea power was more important continujed into the twentieth century, when World War I added a third dimension to the mix: airpower. One lesson of that war seemed to be that if a land power like Germany and a sea power like Britain threw their might against each other, the most likely result would be a stalemate in France, where soldiers slogged it out in trenches for four awful years.

That was where Douhet and Mitchell and the other airpower theorists came in. Not only would airplanes let you leapfrog those trenches; they would also let you destroy your enemy's factories and demoralize his people, cutting the support from under his war plan. Airpower had a new rationale. The trouble is, the evidence suggested that airpower was not the perfect strategy either. After World War II, the Allies discovered that German military production actually rose during most years of the American and British bombing campaigns. And the air war in Desert Storm was not sufficient to force Iraq out of Kuwait.

Even recently, when the United States tried to fight wars almost entirely from the air in Bosnia, Kosovo, and Afghanistan with smart bombs, ground forces, in the form of local surrogates, were critical. In Bosnia, the surrogates were ethnic Croats and Muslims. In Kosovo, the surrogates were ethnic Albanians. And in Afghanistan, the surrogates were mainly ethnic Tajiks, Uzbeks, and Hazaras in the Northern Alliance.

So, by the later years of the twentieth century, we had three competing theories about military power. All seemed to have some validity, but none seemed exceptionally superior to the others. You could take your choice and decide how to get geopolitical leverage: control the land, control the sea, or control the skies. But if anyone was paying close attention, there were already signs that a fourth form of military power might be even more important: control of electronic information systems.

As 1917 opened, World War I had dragged on for two and a half years. The Allies realized they were at the end of their rope. In the East, Russia was about to be overtaken by civil war. The French forces on the Western Front teetered on collapse. Britain faced the prospect of Germany restarting unrestricted submarine warfare. British leaders were desperate for the United States to join in.

The British got the opportunity they needed when they intercepted a coded telegram from German foreign minister Arthur Zimmermann to his representative in Mexico. Zimmermann instructed the German ambassador to inform the Mexican government as follows:

We intend to begin on the first of February unrestricted submarine warfare. We shall endeavor in spite of this to keep the United States of America neutral. In the event of this not succeeding, we make Mexico a proposal or alliance on the following basis: make war together, make peace together, generous financial support and an understanding on our part that Mexico is to reconquer the lost territory in Texas, New Mexico, and Arizona. The settlement in detail is left to you. You will inform the President of the above most secretly as soon as the outbreak of war with the United States of America is certain and add the suggestion that he should, on his own initiative, invite Japan to immediate adherence and at the same time mediate between Japan and ourselves. Please call the President's attention to the fact that the ruthless employment of our submarines now offers the prospect of compelling England in a few months to make peace. Signed, Zimmermann.[6]

The British intercepted the message in January; they passed it to the Americans in February; President Wilson revealed it to the public in March; and Congress declared war on Germany in April. U.S. troops tipped the balance in favor of the Allies. The war was over by November 1918.

The Zimmermann telegram is one of the great stories of espionage. Even today, you will find a copy of the decoded intercept hanging on the wall of briefing rooms at the National Security Agency, like a leopard skin from a big-game hunt, testifying to the power of SIGINT.

How did the British get their opportunity to turn World War I around? When the war began, the British cut the few undersea cable lines that connected Germany with the outside world. After that, the German foreign ministry had only two communication options. They could use radio, which the British could intercept. Or they could use Swedish diplomatic channels, which traveled on lines that passed through Great Britain. The Germans believed that their messages were still safe because they were coded. They did not know that British naval intelligence—the famous cryptographers working in Room 40 of the Admiralty—had broken their code.

Forget about command of the land or command of the sea; the British had *command of the nets*. They had better information technology; they were better at ciphers. But the most important thing was that they had better control over the networks themselves. Most cable lines went through Britain. They could interrupt the few links that passed directly to Germany, and could monitor the traffic through the ones that remained.

This gave the British a better shot at intercepting a really critical message. The codebreaking prowess of Room 40 was important, but the codebreakers would not have had a chance to work their wonders if the British did not have command of the nets.[7]

Sea power didn't eliminate the U-boats for Britain, and control of the Heartland did not enable Germany to beat the British and French armies. But command of the nets, primitive as they were in that age, put Britain ahead of Germany's decision cycle. Before Zimmermann could seal his deal with Mexico, the Brits used the information to bring the United States into the war.

World War II provided a similar lesson. By late 1941 Germany controlled the Heartland, had command of the sea, *and* command of the air. But again the British had something better: command of the nets. Thanks to Ultra, British pilots knew the Luftwaffe's plans to bomb Britain almost as soon as the German pilots did. Getting ahead of the decision cycle meant everything. At Midway, the outnumbered Americans surprised the Japanese thanks to Magic, a masterpiece of U.S. codebreaking.

If the Cold War had become World War III, we might have had a third example. We just don't know who would have come out ahead. As we have seen, the United States tapped into Soviet communications in many places: undersea, underground, in the air, even from space. Would this have enabled us to beat the Soviet Union's decision cycle?

NSA veterans are confident that the United States would have come out ahead. Perhaps they are right. But the Soviet Union had its own cards to play. As it turned out, the Soviets also were fighting for an advantage on the electronic front—only natural considering their interest in *radioelektronaya bor'ba*. Their ultimate weapon may have been John Walker, a Navy petty officer who provided the Soviet KGB with keys to U.S. ciphers beginning in October 1967.

"Do you understand what this means, the significance of this compromise?" asked Boris Aleksandrovich Solomatin rhetorically nearly twenty years later. Solomatin was the KGB's Head of Anti-American Operations. "For more than seventeen years, Walker enabled your enemies to read your most sensitive military secrets. We knew everything!" he says. "There has never been a security breach of this magnitude and length in the history of espionage. Seventeen years we were able to read your cables!"

In effect, the keys Walker provided enabled the Soviets to get inside our decision cycle. Or, as Solomatin put it, "He enabled us to understand your true intentions. It was impossible for you to bluff when we were reading your cables."[8] Who had command of the nets in the Cold War? An interesting question, and really hard to answer—which in itself should give us concern, since the question is even more important today.

Winning command of the nets is much more complex than winning command of the land, sea, and air, partly because it is so hard to tell if you have succeeded. It is a question about "measures of effectiveness," essentially the same question Andy Marshall was working on in the 1950s when he compared Soviet and American nuclear forces.

It is a lot easier to realize that two opposing military forces are asymmetric than to devise a common measure to relate one to another. How do you compare a Soviet-built tank with an American one? How can you measure the effectiveness of an Indian fighter squadron against its Pakistani counterpart? Or—a tough one—how do you measure the power of an Israeli Merkava tank against a Palestinian human bomb?

None of these net assessments is easy, but command of the nets is especially hard to measure because so much of the information technology and services that would be involved in a cyber war belong not to governments but the private sector. Indeed, today the most significant measure of whether a country has command of the nets may be *market share*. Protecting an information system, or getting access to someone else's, is hard enough. But it can be much, much harder if the system is designed, built, or owned by a company in a hostile foreign country.

So one might think that any country that had the leading hardware, software, and online services companies would do its best to keep this lead. One might also think that if protecting and penetrating networks depend more than ever on cooperation between the private sector and government, public officials would do whatever they could to foster cooperation and good will with the IT industry.

Yet, U.S. policy often seemed 180 degrees out of synch from this ideal during much of the 1990s. Take market dominance, for example. Microsoft, Intel, and America Online were all the targets of antitrust suits or investigation by the Department of Justice during the past decade. American IT companies were targeted by the U.S. government precisely because they were *too* successful.

It is hard for government to try to develop a close relationship with information companies with one hand, while trying to break them up with the other.

Yet the issue that really soured relations between the new IT industry and the government was encryption. It all started in 1991, when Philip Zimmermann developed a program he called Pretty Good Privacy, or PGP. Zimmermann's program scrambled messages with an algorithm that was inherently hard to crack. The solution to the scrambling process was linked to factoring, or the mathematics of finding the lowest combination of numbers a number can be divided by. Although some strategies are more efficient than others, factoring is basically a trial-and-error process that requires brute force computing power, so by basing a cipher on factoring, you also make decryption a brute force problem dependent mainly on the amount of computer capacity your opponent can put on the task.

PGP was not only hard to crack; experts could tell you how hard. To factor a number (or conclude that it is prime), one basically has to try out all the different possible combinations of prime numbers that might be divided into it without a remainder. For example, suppose cracking a cipher depended on knowing the factors of a key containing the number 628,933. Even if you had a table of prime numbers from 2 to 1,000 and proceeded sequentially, you would still need to perform 123 division operations before you found that 628,933 could be factored as the product of 677 and 929. Assuming it takes about ten seconds to perform long division on a six-digit number by hand, you know it will require a person about twenty-one minutes to factor the number.

A computer, of course, could do this particular task in a millisecond. But what if the number were much, much longer—say, a block of numbers comprised of hundreds of digits? (The longest known prime number is 2,098,960 digits long; printed in a standard 12-point Times New Roman font, it fills 564 pages.) Since we know the amount of time a computer requires to perform a single division operation, one can extrapolate and estimate the time the computer would need to repeat the process for all the possible combinations of potential factors.

Knowing this, all Zimmermann had to do was choose a key that was so long that even the most powerful computer would need several years to try all the possible combinations of factors. The result was a cipher that was, for all practical purposes, unbreakable.

It hardly made any sense to regulate encryption because the government could not control the dissemination of PGP even if it wanted to. Encryption software is like any other software, and can be sent anywhere in the world over the Internet in seconds. Once Zimmermann invented PGP it was impossible for government officials to eliminate strong encryption; the basic principles were well understood. But that did not keep the feds from trying.

Under the law, exporting PGP was illegal without government approval—federal regulations put it in the same category as tanks and machine guns—and government officials—specifically, the FBI and NSA—made it clear they would not go along. They tried to convince U.S. IT companies to use the government's own commercial-grade cipher, Skipjack, which would be distributed on a microprocessor that came to be called the Clipper Chip. Skipjack contained a "back door" that would allow law enforcement and intelligence agencies to break the code if they obtained the key from a third party under legally prescribed procedures. (This process was called "key escrow.")

American software manufacturers knew that foreigners would be loath to buy, say, accounting software if they believed that the U.S. government could read their files under *any* conditions. They also worried about whether a key escrow system would be secure; otherwise, any slip-ups by the third parties holding the keys could put them out of business. So the IT industry and the government quarreled for several years in the press and congressional hearings. Suspicion and sore feelings between the two remain, even today.

The irony of the encryption issue was that even many law enforcement and intelligence officials believed the government's policy was, well, dumb. They had encountered new technologies that threatened to make their job tougher before—digital microwave communications and fiber optics, to name two. Each time they had worked with industry quietly to get a better understanding of the technology. That's the nature of the SIGINT business. Sources dry up, which is why you are always supposed to be developing new ones.

What is more, it was easy to make a case that the nation as a whole—and, for that matter, American military forces—would benefit if everyone used strong commercial encryption. Strong encryption cannot guarantee that a computer network is secure against a foreign hacker attack, but it is probably impossible to make a network secure without strong encryption. Never mind the banks and utilities; as we have seen, over 90 percent of all *military* communications travel over commercial lines. Without strong encryption, those lines could fail too.

One could tell, though, that these disagreements run deeper than just quibbles over technology and policy details. The recent disputes reflect a clash of cultures. Part of the problem was probably just history. The first-generation computer companies such as IBM, Burrows, Sperry, NCR, Control Data, and Digital were mainly based in the East and the Midwest. So was AT&T, which operated as a heavily regulated government-sanctioned monopoly until its breakup in 1984. Most of these companies had long histories as contractors to the Department of Defense or other government agencies.

These companies were accustomed to cooperating with the government, even when "cooperation" really meant following instructions. They also shared similar cultures. There were many parallels: hierarchical organizations, formal rules, and even uniform dress code. If you were in the Army, you wore green. If you worked for IBM, you wore dark blue.

The new companies that led the personal computer and Internet revolutions—Intel, Apple, Netscape, Oracle, and, of course, Microsoft—were different. Most took root on the West Coast. Many corporate leaders had little experience with government and had never served in the military, having been born too late to be eligible for the Vietnam-era draft. Bill Gates was born in 1955, as was Steve Jobs, co-founder and current CEO of Apple Computer. Steve Case, founder of America Online, was born in 1958, and Marc Andreessen, designer of Mosaic and co-founder of Netscape, was born in 1973.

The new generation learned computers on their own. For that matter, many didn't even bother with a formal education. Gates and Jobs both left college early to concentrate on business; Andreessen completed a normal stay at the University of Illinois, but once claimed he was not sure whether he received a degree or not. The new models for success were the start-up and the IPO, not climbing the corporate ladder; and this new generation believed that the consumer market was more important than government sales.

So, for an entire decade, the government had a paradoxical policy. It failed to pass measures that would encourage the IT industry to make *better* products that would *stengthen* security, like liability law and regulations that would make software harder to hack and more reliable. At the same time, it tried to enforce measures that would encourage the IT industry to make *inferior* products that would *weaken* security, namely in the area of encryption.

To be sure, industry was not blameless either—nor, for that matter, pure. Despite its professed laissez-faire, free market ethos, the IT industry often was as apt to suck down federal dollars as any other industry when the opportunity availed itself. Andreessen himself worked on Mosaic under government-funded research, and Larry Ellison created Oracle partly with air force and CIA funding.[9] And the main reason the software industry has evaded reasonable regulation is that it was so successful in lobbying Washington.

To make matters worse, the government was losing clout. It was no longer the most important customer for computers. Often it did not even have the best technology. For decades, NSA owned the most powerful computers in the world. NSA often gave companies like Cray and CDC the contracts that got them started. With this kind of leadership, the National Institute of Standards and Technology, with NSA working in the background, could issue an encryption standard, and industry would adopt it because there was nothing better. Today the most powerful computers are often in the private sector. Boeing uses them to generate three-dimensional designs for airliners, and Pixar uses them to create animated cartoons.

Finally, in 1999, the feds relented and more or less lifted controls on encryption, but by that time the damage was done. A generation of IT entrepreneurs and developers had come of age believing the government was their adversary. By their thinking, government was just another threat that could put them out of business, in the same league with high interest rates, tightfisted investment bankers, and locusts.

Sometimes the policies that give a country the "information edge" over its adversaries may seem counterintuitive. For proof, recall the Falklands War.

In early April 1982 Argentina invaded and captured the Falkland Islands, an archipelago about 300 miles off its coast. Argentina claimed the islands, but the British had ruled them since colonial times and were not ready to give them up, especially to a military junta using force.

The war could not have come at a worse time or place for Britain. Its economy had declined throughout the 1970s. In 1979 British voters, desperate for a solution to their economic woes and weary of powerful labor unions, had elected a Conservative government—headed by Margaret Thatcher.

Thatcher pushed a free-market economic policy that seemed promising, but which had not yet turned the British economy around. Even in her own party, the "wets" were becoming nervous about Thatcher's economic policies and hard-line stand with the Soviet Union. Labor unions resisted her policies every step. By 1982 her popularity was sagging badly.

It was little wonder the Argentine military junta thought they might be able to shore up their own government by grabbing the Falklands. They were surprised when the British refused to back down, scraped together as many ships and aircraft as they could, and sent a fleet across thousands of miles of ocean to retake the islands, in what they dubbed Operation Corporate.

It took the British fleet eight weeks to make the journey to the South Atlantic. The war stretched both countries to the limit. The British, of course, were fighting a war half a world away from home. However, it was also hard for the Argentines to attack the British. The

Brits bombed the only airfield on the islands (eliminating it as a base for Argentine fighter aircraft), and the Falklands were at the very edge of the range of the aircraft Argentina could fly from the mainland.

In the ensuing battle, Argentine air force pilots attacked the British warships. Argentine pilots scored hit after hit on British ships, but as often as not, their bombs failed to explode. The Argentine pilots were using bombs their country had bought in friendlier times from Britain. The Argentines did not realize the bombs were designed not to explode if dropped at altitudes of less than 200 feet. The British understood exactly what was happening and tailored their tactics accordingly, forcing the Argentines to fly low.[10]

The margin of victory in the Falklands War was incredibly slim. So imagine if some shortsighted bureaucrat in Whitehall had decided a few years earlier that it was a bad idea for Britain to be sending 1,000-pound bombs to a country with outstanding territorial claims against the Crown.

The Argentines might have bought their bombs from someone who would have clued them in on the problem. They would have fixed the fuses or altered their tactics. The bombs that hit would have exploded and the Argentines would have sunk much of the British fleet. Operation Corporate would have failed. Margaret Thatcher would have lost office. Free-market economics, confronting the Soviet Union, and promoting the development of democracy in Latin America would have all lost ground. History would have been changed.

The point is, a country that hopes to dominate technology must be willing to sell technology. Countries that want command of the nets must be willing to sell computers, software, and communications services. Otherwise they will lose the leverage that the seller gains.

Unfortunately, autocratic bureaucrats, often backed by overzealous politicians, believe American companies are run by knaves and fools who would sell out their country if left to their own devices. They think American businessmen are "rope sellers," to use Lenin's turn of phrase—bourgeois merchants who want to make a buck so badly that they would sell the hangman the noose for their own execution.

These officials want tougher controls, and have a missionary's zeal when it comes to implementing the current ones. Almost everyone in the information industries has been affected—the controls on encryption were, as we have seen, export controls—but companies that build communications satellites have been the biggest losers.

Given the importance of maintaining "command of the nets" today, if anyone were planning to build big pipelines in the sky to carry data through space, it would seem that we would want the pipes to be American, and until recently they were. Tougher export controls is one reason foreign companies have made inroads. The problems started in the mid-1990s, when U.S. officials discovered China was making major improvements in the accuracy of their long-range missiles. Some of the Chinese gains seemed to have come from espionage. Some seemed to be homegrown. But some people claimed they had come from U.S. aerospace companies.

Satellite manufacturers like Loral, Hughes, and Lockheed Martin had all used the Long March rocket, manufactured by China's Great Wall Industries. Chinese rockets were cheaper, and a satellite company could save tens of millions of dollars by using a Chinese launch. Alas, Long Marches began to have several launch failures in the early 1990s—that is, they blew up, destroying their multimillion-dollar payloads in the process. Insurance companies wanted Great Wall to do a detailed review to find out why. The American satellite companies that wanted to use the Long March agreed to help in the review.

The U.S. companies did not hide their assistance, since no one thought they were doing anything wrong. In fact, they wanted as much publicity as they could get, because they wanted to convince insurance underwriters worldwide that the Long March was fixed. But when U.S. officials found out, alarm bells went off.

The officials claimed that American engineers had revealed sensitive technologies to the Chinese. Yet a close look at the evidence suggests that the Chinese may well have given away more secrets than they got. For example, one of the components that the Chinese supposedly "improved" with the help of the Americans was their Long March guidance system. It turned out that the electrical connections inside the unit occasionally broke because of the vibrations that passed through the rocket during launch. So one of the companies told Great Wall how to fix it; it was a matter of soldering technique.

According to the story, the Chinese used this know-how in their military missiles. But let's see if we have this straight: The Chinese tell engineers from a U.S. defense contractor—engineers who regularly meet with Pentagon officials—the innermost details about the guidance system used in one of their newest missiles. We tell them how to solder an electrical joint. Sounds like a pretty good deal. Especially when you consider that U.S. aerospace companies solder their joints according to U.S. military specifications . . . which are published by the U.S. Department of Defense . . . and are posted on the Internet.

Didn't matter. Export controls are hard for the public to understand, and easy for politicians and bureaucrats to spin. Many of the critics of the U.S. satellite companies were conservatives, like Representative Christopher Cox (R-Cal.), Senator Jon Kyl (R-Ariz.), and *New York Times* pundit William Safire, who all usually support free trade, free speech, and gaining greater knowledge about the innards of Chinese guidance systems.

Talk to export control officials for more than two minutes, and they will sanctimoniously begin to recite their mantra: "Exporting goods is not a right, but a privilege." Technically, they have a point. They are referring to a legal opinion the Supreme Court issued, *United States v. Curtiss-Wright Export,* back in 1932. The Court ruled that controlling exports was traditionally one of the sovereign right of kings. Since the Constitution does not explicitly limit this power, the Court opined that the President inherited it from the British monarchy. So the export officials are really justifying their authority on the fact that George III could restrict trade.

Of course, George III's inclination to control exports also happened to be one of the reasons the Colonists started the American Revolution.The reality is that export control laws are no different from any other law: Congress passes a bill, the President signs it into law, and civil servants implement it. And that's the real problem. The organizations responsible for licensing exports—the lead agency is the State Department's Office of Defense Trade Controls (ODTC)—have traditionally operated with a level of autonomy that even a Roman prefect would have envied.

Bureaucrats at ODTC have civil service tenure, and until recently elected officials were reluctant to intercede. They feared that if a hostile foreign power ever appeared to benefit from access to U.S. technology they would be accused of being soft on national security. The

Bush administration re-organized the export control office in November 2002; some top officials were effectively demoted when their responsibilities were divided among several new positions.

Even so, export controls remain one of the few instances in which a U.S. citizen can make an honest effort to obey the rules and still wind up on the wrong side of the law. It is easy to make mistakes, because the rules about what constitutes "technical data"—that is, the legal term for controlled information—are broadly defined and open to debate. ODTC does not even have to prove that technical data actually changed hands. Simply talking with a foreign engineer can lead to a charge of a violation. As one export official once told me, "If you put two engineers in a room together, before you know it, technology transfer occurs!"

It remains to be seen whether reorganization will change the notoriously Byzantine export licensing process. But what has been lost in all of this is how, or even whether, export controls really make the United States more secure or less secure. Obviously, no one wants U.S. companies to sell fighter jets, missiles, or nuclear reactors to rogue states. But that is rarely the issue today. Now export controls try to limit "dual use technologies"—often information technologies that are mainly used in commercial industry or consumer goods, but which potentially could be used in a weapon. One case offers some lessons in how misguided this policy can be.

Sometime in 2000 China's Huawei Technologies sold Iraq fiberoptic communications components, which Iraq used to improve its air defense network. This was a violation of the UN embargo on trade with Iraq, but what really made many pundits and politicians in the United States angry was that Huawei had originally obtained the fiberoptic technology from Motorola, an American company. After the press reported the incident, these pundits and politicians cited the Huawei case as evidence of the need for tougher export controls. They seemed to miss a minor point: The original sale wasn't to Iraq; it was to China. Did they want to ban *all* exports of the technology? And if not, how did they expect to prevent Iraq from getting the technology through a third party?

But here's the crunch issue: When U.S. military officials were planning air strikes for a war against Iraq in 2003, which would have been better—for the Iraqis to be equipped with a communications system that was totally foreign to the United States? Or one that American engineers understood in detail, and understood how to neutralize?

When firepower and speed decided the outcome of wars, it made sense to keep powerful, fast weapons out of the hands of our adversaries. Since it was hard for most countries to build tanks, battleships, and bombers, export controls had a chance of working.

Today the ability to command information is almost always more important than firepower and speed, and information technology is so widely available that it is almost always impossible to control. So, as Tom Rona would say, today it is usually better to have insight into—or, even better, control over—your adversary's information technology than to stop him from getting the technology.

By making it harder to penetrate foreign economies and capture market share, most export controls make it harder to win the information war. If the hard-line supporters of export controls had had their way in 1918, the Germans would have owned those undersea cables and all of our friends in Texas, New Mexico, and Arizona would be paying taxes to Mexico City.

Chapter 17

INFORMATION ARMOR

More than six months after Alan Cullison bought his new computer in Kabul, he published a follow-up story with Andrew Higgins, his colleague. Cullison and Higgins had been combing the backwaters of Russia trying to piece together the Chechen connection with Al Qaeda.[1]

The two reporters recounted how Ayman al-Zawahri, head of Egyptian Islamic Jihad, had tried and failed to establish a new base in Chechnya in December 1996. Today, al-Zawahri is well known as a co-conspirator with bin Laden. But in the mid-1990s he was down and out, and pessimistic about his future.

The high point of al-Zawahri's career had been on October 6, 1981, when members of his group assassinated Anwar el-Sadat during a parade commemorating the Yom Kippur War. Although most Westerners would consider the Yom Kippur War a stalemate—perhaps even a military defeat for Egypt—the Egyptian army considered it one of their great triumphs. The Egyptians had crossed the Suez Canal under fire and pushed the Israelis back into the Sinai Peninsula—no mean feat.

Just before a cease-fire was scheduled to take effect, Ariel Sharon had led a counterattack back across the Canal and encircled much of the Egyptian army. Nevertheless, the war had regained Egypt's pride after its crushing defeat in an earlier war with Israel in 1967. More important, the Egyptian army's gains triggered negotiations that eventually led to the recovery of all of the land Egypt lost in 1967, and established Sadat as a major player in the world.

Negotiations with the West, and especially Israel, were the last thing al-Zawahri wanted and that was why he and his followers hated Sadat and, ultimately, plotted his death. When the parade passed the president's review stand, several of the soldiers—actually, plants from al-Zawahri's organization—broke formation and rushed toward Sadat, firing machine guns and throwing grenades. Sadat and twenty others were killed, including four American diplomats.

Despite this spectacular success, al-Zawahri had never been able to broaden his base in Egypt. His group was under constant pressure from the Egyptian government, which infiltrated the organization. By the early 1990s many members of Egyptian Islamic Jihad had been killed in shootouts or thrown in jail (al-Zawahri himself served three years for his role in the assassination).

Chechnya seemed like it might provide al-Zawahri the same kind of refuge that Afghanistan later offered bin Laden. Al-Zawahri had heard that the Russian government had lost control of the mainly Muslim state. Alas, the Russians picked up al-Zawahri and his band almost as soon as they arrived. The terrorists managed to conceal their identities, though, and the Russians deported them after six months. At that point, al-Zawahri threw in the towel, essentially liquidating his group in a merger with Al Qaeda.

Al-Zawahri became a lieutenant to bin Laden, focusing his efforts on recruitment and managing the terror network. He also had an informal assignment: Maintaining good relations among rival factions. So he was in the loop with the leadership, exchanging memos and notes with other top members—including Mohammed Ataf, the previous owner of Alan Cullison's computer. In one letter recovered from the hard drive, dated May 31, 2001, al-Zawahri tried to get members of the terror network to stop squabbling over money matters.

In retrospect, though, what was most interesting was why he said settling these issues was so important. "Stop digging problems from the grave," he implored; bin Laden had a "project" that required their support.

"Our friend has been successful and is seriously preparing for other successful jobs," wrote al-Zawahri, referring to bin Laden. "Gathering together is a pillar for our success." The plotters of the September 11 attacks had already finished flight training and were well into their analysis of targets and air schedules.

During the summer of 2002, a panel working under the direction of the House and Senate Intelligence Committees investigated intelligence failures leading up to September 11. When Representative Saxby Chambliss presented the results at a press briefing, he said, "This was such a closely held, compartmentalized act of devastation that was carried out by the terrorist community that we don't know of any way it could have been prevented."[2]

Culllison's computer said otherwise. Other correspondence on the hard drive made oblique references to aircraft, the sky, and terrifying events to come. All intelligence is fragmentary, and an essential skill in intelligence is using one piece of evidence to stir up additional pieces of the puzzle. This fact was undeniable: At least some people outside the terrorist cell carrying out the September 11 attack knew about the plot, and even more people had heard about the plan indirectly. There was information out there that the United States did not have, and which would have provided additional clues of the attack—*if* we had made the effort and taken the risks that were required to collect it.

What other information was out there that we did not have? *That* was the question that needed to be asked. Most people wondering how U.S. intelligence failed seemed to be looking for the smoking gun—the single piece of information, which, if it had just reached the right person at the right time, would have sounded the alert. Yet the problem was not just that U.S. intelligence did not pull together all the relevant information. The bigger problem was that U.S. intelligence had not collected all the information it needed. U.S. intelligence had failed because we did not have command of the nets.

The annual report from Lucent to its stockholders illustrates the challenge U.S. intelligence faces today: Every minute, 5 million e-mail messages are sent. Every hour, 35 million voice mail messages are left. Every day 50,000 people sign up for wireless phone service. Every day 37 million people log on to the Internet. Every week 630,000 phone lines are installed. Every 100 days Internet traffic doubles.

U.S. intelligence is drowning in digital data, and this affects every part of the intelligence process. During the Cold War it was easier to detect a signal from Moscow instructing its forces to attack. The potential recipients were well known and hard to miss—like the Soviets' Eighth Guards Army just across the border in East Germany. Analysts knew what kinds of signals to look for (the Eighth Guards Army and the rest of the vast Soviet military establishment practiced invading Western Europe every year) and there were only a few communications links to watch.

Today the problem is much harder. Not only are there more potential recipients to monitor and more data to collect, but the data is often more dispersed and harder to get to. Often the only way to collect digitized communications operating over glass fiber networks, local networks, or cellular systems is with taps near the source. Also, because encryption technology is so widely available, the nearer one gets to the point of transmission *before* the message is encrypted, the better the chances of deciphering it.

In SIGINT jargon, this requirement is called "close access." In many cases, it means that breaking into an opponent's information system requires getting a human being so close to a computer, a server, or a fiberoptic cable that he can touch it. In short, you need a cooperative operator of the equipment that carries the data—or a spy. So why don't we have more spies, or to use the technical term, "HUMINT"— intelligence provided by humans?

After the September 11 attacks, Stansfield Turner, Director of Central Intelligence (DCI) under Jimmy Carter, was a frequent target for blame. Turner took office just after the end of the Vietnam War. With the war over, the CIA had more staff than it needed, and Turner fired hundreds of CIA officers. Ever since, it has become folklore that Turner gutted U.S. HUMINT capabilities—and that is why the CIA has been slipping today. For example, shortly after the September 11 attacks, Tom Clancy, former insurance broker/current novelist/frequent defense commentator, offered this analysis on terrorism in *The O'Reilly Factor*:

> The CIA's Directorate of Operations, which generates what we call human intelligence, and the DO, the Director of Operations,

they're the actual field spooks who actually go out into the field and talk to people and get human intelligence information, they were gutted. First, as a result of the Church Committee hearings and secondly by Stansfield Turner, who was appointed by Jimmy Carter in the late '70s. That capability . . . has never been reestablished.[3]

This explanation overlooks a minor historical detail: It's been twenty-five years—*a quarter century*—since Turner ran the CIA. Four Presidents, twelve Congresses, and six directors of the CIA have come and gone. An intelligence officer could have served out his entire career in the CIA during that time. So if the United States has not been investing enough in HUMINT, we really need to stop blaming Stansfield Turner.

Another explanation for the lack of HUMINT (and, by extension, the September 11 failure) was that the CIA has been overregulated. CIA officers were supposedly prohibited (or at least discouraged) from putting human-rights violators on the U.S. payroll. John Deutch, DCI during Bill Clinton's first term, was the main target of these charges. When Deutch took office, the CIA was ensnared in a scandal. The Agency had recruited Guatemalan army officers who, in addition to providing information to the CIA, were torturing political prisoners.

Under Deutch, many agents with questionable backgrounds were dropped, and the CIA adopted stricter rules for approving agents who had questionable records.[4] Some critics claimed these regulations kept the CIA from recruiting murderers and torturers—in other words, the kind of people one would find in the inner circles of a terrorist cell.

There are many, many reasons to criticize John Deutch and his tenure as Director of Central Intelligence, but in this case he was getting more than his share of the blame. Most members of Congress and top CIA officials on the scene today were there when Deutch was DCI. Most supported the restrictions on controversial HUMINT sources—and on the intelligence budget cuts of the 1990s, and the intelligence priorities that supposedly slighted HUMINT.

Besides, as in the case of Stansfield Turner, the CIA had plenty of time to change any misguided policies after Deutch left office. Despite the oft-repeated line that "HUMINT sources take decades to re-build," that simply is not true.

When William Casey was directing HUMINT operations in Europe for the Office of Strategic Services in World War II, he built a network of agents in just eighteen months.[5] Casey was simply more aggressive and willing to take more chances than the CIA seems willing to accept today. Casey might take, say, one agent with questionable motives or bona fides to get four good ones. Casey and the OSS accepted the risk of bad information or penetration by a mole because we were at war, and if we operated under similar assumptions today, we might get similar results.

The real reasons why we lack the spies needed for "close access" SIGINT are more basic. The CIA's Directorate of Operations as currently configured is just ill suited to collect the kinds of agents we need today. And, like any fifty-year-old bureaucracy, it has proved remarkably resistant to change.

Reuel Marc Gerecht, a writer who has identified himself as a former CIA officer, has written articles that offer some light on the problem.[6] Gerecht describes a CIA that is dysfunctional, tired, aimless, and bureaucratic. It does not reward officers who live under deep cover. In fact, according to Gerecht, many CIA case officers cannot speak the language of the countries they cover (and, hence, of the agents they are supposed to recruit). As a result:

> A former senior Near East Division operative says, "The CIA probably doesn't have a single truly qualified Arabic-speaking officer of Middle Eastern background who can play a believable Muslim fundamentalist who would volunteer to spend years of his life . . . in the mountains of Afghanistan." . . . A younger case officer boils the problem down even further: "Operations that include diarrhea as a way of life don't happen."

Gerecht published his critique just weeks before the September 11 attack—he was mainly reacting to the failure by U.S. intelligence to anticipate the suicide bombing attack nine months earlier against the U.S.S *Cole.* He warned that U.S. intelligence was too complacent. But he was also making a more telling point: Even with more money and fewer restrictions on recruiting agents, the CIA would still lack the skills and opportunities it needed to penetrate information systems.

Margot Williams has a quick smile, oversized glasses, long brown hair, and generally looks as though she might have stepped from the pages of the J. Crew catalogue. It was early on a spring afternoon, so the *Washington Post* newsroom was half empty. Most of the reporters were still out on their beats and would not be back until later to write their stories.

Williams was in the newsroom because she doesn't go out much on most days anyway. She is a researcher at *The Washington Post* and specializes in computer-based reporting—finding and piecing together public records and other information that one can find on the Internet.

Most people are probably unaware of how much information about them is really available, and how easy it is to assemble it with a good search engine, interlinked online databases, and computers. For the sake of illustration, I gave Williams a name to look up. Call him Ron. I had dinner with him a few nights before. "What can you tell me about him?"

She typed and moved the mouse around. Within thirty seconds she gave me his job: "He works for the Central Intelligence Agency." Actually, Ron spent a good portion of his career undercover and is now one of the Agency's top officials responsible for clandestine operations. He's overt now, so we weren't compromising any secrets, but you'll see the problem in a minute.

Williams worked away at her computer, and went to Autotrack.com, a Web site specializing in public record searches. In the next five minutes Williams showed me how she could find out where Ron had lived during the past fifteen years, his wife and children, his ex-wife, all court proceedings that he might have been in, and all properties that he owned. Williams also listed all of Ron's previous places of employment, where he went to high school and college, his place of birth, date of birth, and Social Security number (these last two items, incidentally, are what are used to pass security clearances).

"If we wanted to, we could go back and interview all of his old classmates," she said. That would take some legwork, but it would have fleshed out his background.

Once Williams figured out that Ron was a CIA official, she then pulled all of the files in *The Washington Post*'s own archives that mentioned him. By this time we had dozens of records and documents on an official whose relationship to the Agency was, at one time, officially secret.

"We can also get his credit rating if you want," Williams said. I declined, mainly because I did not want to pay the $20 the service required. Besides, I really didn't feel compelled to know how much Ron might have had outstanding on his Visa card.

Then, as a final flourish, Williams moved the cursor on her monitor to an icon and clicked the mouse. The icon launched a link analysis tool. Link analysis tools are a distant cousin of relational databases, and yet another example of how software evolves and how different programs are related to each other.

Until about thirty years ago, most databases were hierarchical. Each datum plugged into a tree-and-branch organizational structure similar to an org chart for a company. For example, a soldier might be entered into the database as Ammunition Bearer No. 2, in Platoon A, which is part of Company C, which is part of the 2nd Battalion of the 82nd Airborne Division.

In 1969 an IBM technician, Ted Codd, published a paper proposing a different approach: a relational database.[7] Instead of putting every entry into a pre-defined slot, a relational database tags each datum according to several variables. Each soldier might be described according to, say, height, weight, sex, IQ, academic degrees, and professional certification.

This may seem like a minor twist, but it makes a huge difference in your ability to use a database. With a relational database, you can sort and group entries by whatever variables you can tag in the data that happen to interest you at the moment. Data acrobatics of this kind were impossible when everything was on paper, but once computers became inexpensive, it was easy to put records on magnetic tape or on disks, and do the shuffling electronically.

After Codd came up with the idea, programmers raced to see who could create the best database package and capture the market. Larry

Ellison won the race in the mid-1970s, who, as we have seen, founded a new company with the assistance of contracts from the Central Intelligence Agency.[8] The CIA project was called Oracle.

Ellison took the name for his own software company, and today Oracle is second in size only to Microsoft among software manufacturers. Companies use Oracle software and its cousins to keep track of their inventory and accounting. So, when e-commerce and computerized Just In Time manufacturing took off in the 1990s, so did Oracle.

A link analysis tool works like a relational database that runs in reverse. Instead of finding a single item that fits a list of parameters, it identifies all the parameters that connect several items. A link analysis tool can take a huge collection of documents, phone listings, schedules, and address books, select names from each, and depict visually how each person is related.

So when Williams clicked the icon, the screen went black for an instant, and then, like a flower blossoming, a diagram resembling an intricate web opened up to fill the screen with Ron, the CIA official, at the center. It was a social network, connecting the data in the dozens of documents Williams had found. The program had searched all of the documents and sorted out who knew whom, who owned what, and who belonged to which organizations.

We agreed that Ron was so high up in the CIA that he might be too easy a test case. So we chose another CIA officer from the Directorate of Operations—call him Jose. He was listed both as a CIA officer and, from some old databases, as an official in the State Department.

Jose hadn't worked undercover for years, but Williams showed me how she could go back to identify some of the places the official databases had him listed as a U.S. official posted abroad. She also showed how you could cross-check other databases to find people he might have met—in other words, people who might now be suspected by their own governments as spies working for the United States.

Also, by comparing Ron and Jose's publicly acknowledged employment records, Williams showed me another trick: We could cross-reference several officers who had previously worked undercover, and put together a profile of what current CIA officers working undercover would look like.

For just a brief instant, I thought of giving Williams a couple of names of officers who were still undercover to see what she could do, but quickly realized that was a profoundly bad idea. She had made her point. With enough time Margot, from the comfort of her cubicle on 15th Street, could blow the cover of most anyone using the methods the CIA has traditionally used to hide the secrets of its case officers.

Just to be clear: No one was giving away anything. Every intelligence service in the world that's worth anything knows all about these tricks. The story is just supposed to make a point. Creating cover is a huge problem for the United States today. Most CIA case officers working abroad operate under "official cover"—they pose as U.S. officials, and the only thing that is secret is that they are collecting intelligence. We are fooling no one, except possibly ourselves, by believing hostile governments and organizations that mean us harm do not know who our CIA officers are.

It all goes back to the spiritual godfather of all case officers, Allen Dulles, who ran the U.S. station in Bern for the Office of Strategic Services during World War II. Switzerland was neutral and lay astride the diplomatic and financial crossroads of Europe leading north and south, and east and west. So Dulles did not have to find and recruit German generals, Italian colonels, or Austrian businessmen. They came to him. Most of Dulles's agents were walk-ins.

Because it lacks truly effective cover, the CIA's Directorate of Operations today often operates less like a hard-core intelligence organization and more like a back-channel State Department, through which the United States can conduct sensitive relations with foreign governments, officials, and organizations on a secret or not-for-attribution basis. In fact, the signature of a case officer is so distinctive that putting people under official cover makes it *more* likely they will be detected. When we give the officer's credentials to the host government, we are telling our opponents whom to watch.

To be sure, the semi-secret case officer works fine when the CIA needs to cooperate with (or recruit) foreign military, intelligence, and law enforcement officials. But this approach is bumping up against its limits in the war against new threats. We need HUMINT today to recruit the sysop at a Middle Eastern Internet service provider, or the janitor at a telephone company in the Far East. Why would an official from the U.S. government be meeting with a sysop? And how much credit do you suppose a CIA officer receives for recruiting the janitor who cleans the computer room at a telephone company?[9]

Alas, the case officer model is deeply ingrained in the CIA organizational structure, personnel system, and tradecraft. Suppose you are an American patriot who happens to be dark-skinned, fluent in Farsi, and eager to serve your country. The current system gives you two options.

Choose Option A—working under "non-official cover" and effectively you must work two jobs: as, say, a manager in a company overseas at a starting salary of $70,000 to $90,000 a year, and as an intelligence officer working for the U.S. government at a starting salary of $45,000 to $60,000 a year. Except that you will get only the lower salary. You have to spend most of the time abroad. You won't be able to meet with most of your real colleagues. And, by the way, if you get caught, you will get thrown in jail (or worse) in a backwater country that is not exactly up to date in its application of the rule of law.

In Option B, you get to live in McLean, Virginia, and drive fifteen minutes to work. You get to see your buddies in the cafeteria during lunch, and you can send your kids to Fairfax County schools. You can network like crazy, get steady promotions, and every couple of years go abroad and pose as a U.S. official.

With this kind of choice, it's a wonder we can get anyone at all to work under deep cover. But that's exactly what we need if we plan to crack many of the most critical information networks.

The problem parallels a principle in economics—Gresham's law, named after an English banker of the 1500s, Sir Thomas Gresham. Its simplest form is "Bad money drives out good money." For example, if some $1 coins were made of silver and others were made of tin, soon only the tin coins would remain in circulation. People would keep the silver coins because silver is harder to come by and the tin ones have the same official value.

The same thing happens in organizations that try to operate officers under deep cover and official cover. Soon almost all the officers are under official cover. Like tin, official cover is easier to come by, and if the organization rewards all of its officers equally, soon no one will want to operate under deep cover. That is why truly clandestine operations need their own separate organizations and career paths.

In fact, today weak cover can be positively dangerous to anyone trying to work under deep cover. Because it is so easy to spot intelligence officers working under official cover, they are the *last* people anyone who is serious about hiding their links to the U.S. government wants to meet. Such an encounter would beg the question of why someone who is supposedly a normal, unassuming private party is meeting with a person known to be working undercover for the U.S. government.

After September 11 there were lots of reports in the media saying that the CIA was flooded with applications and that the Directorate of Operations was taking in more recruits. It would be interesting for a Senator or Congressman on one of the intelligence oversight committees to ask whether the percentage of intelligence officers who are under non-official cover has gone up.

One CIA official (who realizes that official cover is just pretend cover) observes, "Cover is like soap; once you rinse it off, you can't put it back on." Effective cover is only going to get tougher when biometrics makes it possible to follow people around as they assume different identities. Once you identify someone as CIA, they are marked for life. Even worse, anyone they contact will become suspect. So cover has to be a way of life.

It's an irony of the Information Age: In an era of easily available, rapidly transportable information, often the most valuable—and scarcest—asset you can have is cover: the ability to obscure, hide, or withhold information. This is a new form of warfare, but we have experience that should tell us what works and what doesn't. Consider the story of the Cow-Killer Rocket.

In 1960 the Naval Research Laboratory developed a new scientific satellite it called SOLRAD (for "SOLar RADiation" experiment). NRL scientists had been exploring the upper atmosphere and the fringes of space for more than fifteen years. In fact, the lab had developed the first large U.S. rocket, the Viking, which is the direct ancestor of all rockets the United States builds today.

SOLRAD was a twenty-inch, 250-pound aluminum ball that mapped variations in solar energy at various latitudes. This is useful, for example, for predicting locations where one might have trouble communicating by shortwave radio, which uses the ionosphere as a reflector and varies with fluctuations in solar activity. Sometimes solar activity can also create static.

The Navy launched SOLRAD satellites in a two-for-one deal where SOLRAD was mounted on top of a larger Navy satellite, Transit, a predecessor of today's Global Positioning System. The satellites had to fly in a north-south orbit, so the Navy used a Thor rocket—a converted Air Force ballistic missile—to launch the two satellites from Cape Canaveral to the south, down the Florida coast. The rocket's trajectory just nicked the eastern end of Cuba. Once in orbit, SOLRAD would pop off, leave the Transit satellite, and go into a higher orbit.

The Navy launched its first SOLRAD-Transit combo in March 1960, and everything worked as planned. Then in November the Navy tried its second mission in the series, and everything went wrong.

Something broke on the rocket, the motor cut out early, and the Thor began to fall back to earth. Ordinarily this was no big deal. Rockets failed a lot back then, and all American boosters carry destruct charges. So if one begins to go off course, the range safety officer can press a button and—kabam!—the debris falls harmlessly into the ocean.

Except this time. It may have been the first recorded case in the American space program when a destruct charge failed to, well, destruct. The Thor stayed more or less intact, and instead of sending a scientific satellite into the frontiers of space, the United States was suddenly launching an intermediate-range ballistic missile directly at the Republic of Cuba.

The Thor—complete with SOLRAD and Transit—ended its abbreviated mission on a farm in the Cuban countryside. If you believed the local authorities, an innocent bovine in the province of Oriente was in the way, got hit by the debris, and met her Maker in cow heaven. A few days later 300 students, several cows, and a bull marched on the U.S. embassy in Havana. The Cuban press began to carry stories of "Yankee provocations," and the Cuban government sent an official protest to the United Nations.

American officials wanted to bury this incident—and fast. It may have been that the United States had technically violated Cuban airspace. It may have been that the United States was in a high-profile competition with the Soviet Union, and any slip-up in the space race was embarrassing. Or it may have been the fact that at that particular moment the CIA had 1,300 Cuban exiles training in Guatemala for an invasion at the Bay of Pigs just five months later.

Whatever the case, U.S. authorities agreed to pay Cuba for "damages"—presumably, this was Elsie—and promised never to repeat the performance again. The cow was given a state funeral as a victim of imperial aggression, and in the space community the tale of the Cow Killer Rocket became an oft-cited lesson in the risks of launching satellites into orbit.[10]

Then again, the Americans may have had another reason for wanting to defuse the incident quickly. SOLRAD was a real scientific experiment that produced real scientific data. But the satellite also contained a radio receiver that could secretly detect radar transmitters. The classified name for the project was GRAB ("Galactic Radiation And Background"), and SOLRAD was actually the cover for the first U.S. reconnaissance satellite.[11] As GRAB passed over the Soviet Union, it transmitted signals to a Navy ship offshore, and the National Security Agency used the data to plot the location of Soviet air defenses.

Even after SOLRAD crashed into Cuba, no one found out about GRAB. The GRAB program remained secret until 1998, when the Navy decided to declassify the project to celebrate the Naval Research Laboratory's seventy-fifth anniversary.[12] The Navy was, in fact, so successful in keeping GRAB undercover that, in 1995 when the CIA published a history of CORONA, its own early satellite project, it inadvertently referred to it as America's "first successful reconnaissance satellite," even though GRAB was operational five months earlier. The program was still secret, even to many CIA analysts.[13]

CORONA, incidentally, was also a very successful program. CORONA was the first satellite to take reconnaissance photos from space. But CORONA was also a step-by-step lesson in how not to create a cover story for a secret project.

The CORONA project was a spin-off from a larger Air Force program called WS-117L, an effort to develop several space technologies for the military. The imagery part of the project was too important to allow to get bogged down, so the Eisenhower administration decided to separate it and gave it to the CIA for fast-track deployment. The CIA planned to launch CORONA satellites from Vandenberg Air Force Base in California. The satellites were to take pictures for a month or two, and then drop a capsule containing the film back to earth, where an airplane would snag it out of the air by hooking its parachute in a giant trapezelike device.

The CIA, to be frank, never really thought carefully about how it was going to explain launching a satellite into orbit and recovering it after re-entry every second or third month. The CIA finally came up with the idea of saying CORONA was a space biology experiment. According to the story, the United States planned to put black mice of the C-57 strain into orbit and recover them from space for study. The cover program was called DISCOVERER.[14]

The story was at least plausible. The Soviets had launched Laika, a canine mainly of Siberian husky heritage, into orbit on Sputnik II, and the United States was training a chimpanzee astronaut corps that would try out the Mercury capsule before NASA moved the next rung up the primate ladder. Alas, the CIA experiment in space biology was jinxed from the start.

The CIA thought—well, hoped—that some Thors would come along with their portion of the W-117L program when they took it over from the Air Force. But the Air Force declined. So the CIA next went to the Defense Advanced Research Projects Agency to see if they would fund the DISCOVERER launch, only to be told that sending rodents into orbit was not exactly a high priority for DARPA. Eventually the CIA had to go directly to the White House to get the money.

All this was happening as the CIA was struggling just to get the real CORONA missions into orbit, never mind make the satellite work. But the CIA bit the bullet, and scheduled DISCOVERER III to carry four black mice into space. The lead crew—four mice, seven to ten weeks old, just over one ounce apiece—trained with sixty other mice at the Air Force's Aeromedical Field Laboratory at Holloman Air Force Base.

So, on July 20, 1960, the first CIA space mice prepared for their rendezvous with destiny. The countdown was well underway, when someone noticed that the telemetry monitoring the rodents' heartbeats had flat-lined. Scrubbing the mission, the launch team opened the capsule, only to discover that the crew had eaten the Krylon coating inside; it had been designed for a film capsule, and had not been tested to determine whether it was mouse-safe (or mouse-tasty).

After giving the mice the last rites, the launch team sprayed the capsule with a safer material and prepared the backup crew. This time, after a normal countdown, the Thor took off, turned south, and proceeded to arc into—not over, into—the Pacific Ocean.

At this point the CIA simply dropped the cover story and adopted the strategy of just having the Air Force tell everyone the purpose of the DISCOVERER flights was secret (the name CORONA was itself classified). Even so, everyone who cared knew exactly what the launches were. After the first successful launch—DISCOVERER XIII—was recovered, *The New York Times* reported that "the same ejection and recovery techniques eventually will be used for returning photographs taken by reconnaissance satellites."[15]

The lesson: You are better off finding cover than creating it. It helps to have a big organization already operating a program, like NRL's upper atmospheric research program, that you can embed a secret project in. The best situation is often when you can dovetail the secret operation as a quiet part of a visible—even highly visible—operation.

That was exactly how the Navy hid another secret project twenty years later.

When tens of millions of moviegoers worldwide went to see Leonardo DiCaprio and Kate Winslet in James Cameron's 1997 film *Titanic*, few of them realized that they were seeing the end result of a highly classified Navy program. Robert Ballard had originally found the *Titanic* in 1985 as part of a program that was, in effect, a cover for a secret mission to find two submarines the Navy lost in the 1960s.

One was the U.S.S. *Thresher*, which sank 200 miles off the coast of Cape Cod on April 10, 1963. Apparently a pipe open to the sea burst. The submarine took on water, the electrical controls to her nuclear reactor shorted out, and the engines shut down. As the *Thresher* took on more water, she sank deeper, until she reached her crush depth. The pressure of the surrounding ocean imploded her like a bomb. The debris—mainly shreds and fragments—settled on the sea bottom, 8,600 feet down. All 129 crew on board were lost. It was the Navy's worst submarine accident.

Four years later, in May 1967, the Navy lost another submarine, the U.S.S. *Scorpion*. Even today some experts still disagree about what happened, but most believe a torpedo on board malfunctioned. The torpedo either exploded, breaching a hole in the hull, or circled back on the submarine when the captain jettisoned the weapon. In any event, the submarine broke apart and sank in two pieces, plummeting to a depth of more than 10,000 feet in the broad Atlantic, 400 miles southwest of the Azores. All 99 men on board died.

The Navy had located both wrecks in the 1960s, and had sent a few men down in a bathyscaphe—a deep-sea submersible—to retrieve pieces of the wreckage. But bathyscaphes are slow, unwieldy vehicles, so no one had been able to do a detailed survey of the two wreck sites.

This was a continuing worry for the Navy, which was concerned about the condition of the nuclear reactors on the two subs, and especially the nuclear-armed torpedoes that the *Scorpion* was carrying at the time it sank. However, the Navy did not want to draw attention to the wrecks, lest anyone—like the Soviet Union—get the idea to grab parts of the submarines for themselves. And the Navy has always been sensitive about disclosing its undersea activities. So any survey had to be secret.

Robert Ballard worked for Woods Hole Oceanographic Institute, a research center in Massachusetts. Woods Hole was founded in 1930 by a grant from the Rockefeller Foundation, and today it is an independent, not-for-profit institution, employing about 600 people who work in fields ranging from ocean physics, biology, and engineering to policy studies.[16]

Woods Hole gets its funding from a mix of federal grants, private contributions, and income from an endowment. One of the main grantors was the Office of Naval Research, which began funding research there in World War II. Oceanography is as important to the Navy as, say, topography is to the army and aeronautics to the Air Force, so the funding link is a natural one.

In the early 1980s, scientists in ocean research, like the rest of society, began to appreciate the potential of the new information technology that was beginning to appear. For oceanography—and for sensitive military and intelligence operations at sea—the word was *robotics*.

Navies and the oceanographic research community had used towed sensors for many years. U.S. frigates and submarines had towed sonar devices—their "tails"—off their sterns to detect submarines since the 1960s. The Navy had also developed some simple "sleds" that could photograph and grab objects off the sea bottom. One of the earliest, the Cable Controlled Undersea Recovery Vehicle (CURV) enabled the Navy to retrieve a hydrogen bomb that had been lost off the coast of Palomares, Spain, in 1966 when a B-52 collided with a tanker while refueling and crashed.

Inevitably, the Navy used the technology for intelligence missions, and this boosted the state of the art further. By the late 1960s the Navy modified the nuclear submarine U.S.S. *Halibut* to carry a "fish" that it could release underwater from a compartment in its hull. In some missions, the fish was used to photograph objects of interest, like a sunken Soviet submarine.[17]

Then in 1973 Edwin Link came up with a new idea. Link, a self-taught engineer, had spent the 1920s barnstorming with former members of the Lafayette Escadrille and hanging out with Orville Wright. He first became famous for designing one of the first flight simulators, the Link Trainer, in which thousands of pilots learned to fly during World War II.

After the war, Link got interested in underwater exploration. With the support of J. Seward Johnson, the founder of the Harbor Branch Oceanographic Institute, he designed a series of small submersibles he called Johnson-Sea-Links. The submersibles had a unique design. The front of the submersible was an acrylic bubble holding two operators, who enjoyed a panoramic view. The back was an aluminum chamber with a hatch that allowed two divers to go outside. Chemical scrubbers in each chamber removed carbon dioxide from the air so the crew and divers could breathe.[18]

On June 17, 1973, Link's son, Clayton, was diving with three others to explore the wreck of an old destroyer off Key West. The submersible got caught in some cables. The two operators in the forward compartment survived, but the temperature dropped precipitously in the back, where Clayton and another diver, Albert Stover, rode. The scrubber failed in the cold, and Link and Stover died of carbon dioxide poisoning.

Ed Link was devastated, but focused his grief on designing a device that would rescue divers if a submersible found itself in a similar predicament. Link's solution was to design a two-part underwater robot. The larger section provided a stable base and carried most of the gear. A smaller section, housing lights, a camera, and a claw, could detach and maneuver on the end of a cable. Link called the robot CORD (Cable-Operated Remote Device). A submersible could take CORD with it on a dive and, if the divers got caught like the crew of the Johnson-Sea-Link, use the CORD to remove whatever was in the way.[19]

Link's idea combined the stability and strength of a large robot with the agility of a small one. Ballard took this basic idea and pushed it further. Ballard had taken a year off from Woods Hole and spent some time at Stanford University. Stanford, being located in Silicon Valley, was in the middle of the Information Revolution, and Ballard got the idea of applying the technology to create a new field: virtual exploration. The new sensors, fiberoptic cables, and microprocessors would let an operator "see" the ocean floor from the comfort of a control booth, and use manipulator arms that could handle objects with a deftness that had been impossible before.

Armed with this new technology, Ballard—as much an adventurer as a scientist—could pursue his dream: finding the *Titanic*. People knew where other lost ocean liners—*Andrea Doria, Lusitania, Britannic*—had gone down, and divers had explored most of them. But no one really knew the exact location of the wreck of the *Titanic*. Even if they did, the ocean was so deep where she sank that no one could reach her. All of this, combined with the Edwardian-era hubris that surrounded her loss and the celebrities who died aboard her, made the *Titanic* the grand prize of ocean explorers—almost mythological.

The problem now was funding the technology and the adventure. No one, let alone the U.S. government, was prepared to spend the millions of dollars that would be needed just to find a wrecked liner. But Ballard, in addition to being a scientist and an adventurer, was also an entrepreneur. He proposed a deal to the Office of Naval Research: fund most of the development for a deep-sea robot. Ballard would raise the remaining money. Woods Hole would look for the *Thresher* and *Scorpion* and survey the wreck sites. Then Ballard could use whatever time he had remaining on the way home to look for the *Titanic*.[20]

"They wanted me to apply my skill set to their problem, and I wanted to use their problem to expand my skill set," he recalls. The robot became the ARGO/JASON system. Like CORD, it consisted of two parts. ARGO was the large partner in the combo—about the size of a car—and served as a docking station for JASON, a smaller sub-robot that carried the lights and camera. JASON could reach tight quarters, like the interior of the *Titanic*—or the torpedo room of a submarine that had been lost sixteen years before.

For the Navy, using Ballard's research as a cover offered many advantages; none the least was Ballard himself. Like that other submariner, George Dixon, Ballard had started his military career in the Army, as an ROTC cadet at the University of California, Santa Barbara. "I had three honorable discharges," he says, laughing. Knowing of Ballard's interest in oceanography, during the Vietnam era the government retired him from the land forces, commissioned him into the naval forces, and sent him to Woods Hole as the liaison officer for the Office of Naval Research. When he served out his commitment, he resigned his commission and was eventually tenured as a member of the Woods Hole research staff.[21]

"During the Reagan-era ramp-up, they pulled me back in," he says as he recalls the missions he performed during the 1980s. Ballard had remained an officer in the Naval Reserve. So, whenever the Navy needed him for an operation, Robert D. Ballard, Ph.D., simply became Commander R. D. Ballard, USNR, a bona fide U.S. naval officer performing a secret mission for his country while appearing to the rest of world to be an oceanographer exploring the seas—which he was. The quest for the *Titanic* (and, later, a dozen or so other ocean liners and warships) became stories about Ballard, chronicled in a series of television documentaries and personal memoirs with exquisite illustrations of the lost ships he had found.

It was a great game of three-card monte—the more the public kept its eye on Ballard and his adventures, the less interested it was in the other activities the Woods Hole team was doing along the way. Even when William Broad, a reporter from *The New York Times*, guessed some of the objectives of the program, the public kept focused on the search for lost liners, and no specific operations were revealed.[22]

By publicizing the search for the *Titanic*, which was bound to get the spotlight anyway, and just being coy about how much time Ballard was actually at sea, the Navy succeeded in making sure no one brought up the subject of the *Thresher* and the *Scorpion*. The secret parts of the mission remained secret until Navy officials decided to reveal the survey operation in 1996 to reassure the public that the reactors posed no safety hazards.

In effect, the Navy had used the same approach to collect secret information from the sea as it had to collect secret information from space: Find a large research institute with enough routine activities that a secret project can get lost in the noise. Create a project with a plausible, bona fide scientific function. Blend the sensitive operations into the larger project. Publicize the public operations to the max, and just be nonchalant when the press follows its own instincts for a good story. Simply say nothing about the sensitive part. And be sure you have partners with a real stake in the success of the operation.*

This was exactly the opposite of what the CIA did in the case of CORONA, and, later, in its own deep-sea recovery project, JEN-NIFER. The CIA tried to raise an entire Soviet submarine using a ship christened the *Glomar Explorer*. Again, the program was conceived thinking that you could create cover, rather than use cover that already existed.

The CIA built the 600-foot-long *Glomar Explorer* in 1973. It then tried to pass it off as an ocean mining platform operated by billionaire Howard Hughes. The program—which required thousands of people to be cleared—ran for just over a year before its true purpose was blown (at a cost of tens of millions of dollars) by the *Los Angeles Times* in 1975.[23]

The lesson again: You are more likely to keep a secret program secret if you piggyback it on top of a real program, preferably one that al-

---

* The Navy had reason to be concerned about offering any clues about the location of the submarines or how one might reach them. Russia has its counterpart to the Woods Hole Oceanographic Institute, the Pyotr P. Shirshov Institute of Oceanology. It operated the research ship *Academician Mstislav Keldysh* (counterpart to Woods Hole's *Atlantis II*) and the *Mir* submersibles (counterpart to Woods Hole's *Alvin*). More recently, the Shirshov Institute (now funded by the Russian Academy of Science) used the *Keldysh* to survey the wreck of the Russian nuclear submarine *Kursk*, which was lost in an accident in August 2000, much as Ballard surveyed the *Thresher* and *Scorpion* sites. James Cameron used the *Keldysh* and *Mir* submersibles to film the underwater scenes in *Titanic*.

ready exists. The people working undercover do not live a separate life; they do their secret work as an adjunct to their normal activities. In the trade, this is called "living one's cover."

Instead of hoping case officers can develop access to Information Age targets, we will likely have more success if we find people who already have access—say, the thirty-five-year-old American businessman working abroad who was a Marine back in the 1980s and wouldn't mind spending part of his time working on military or intelligence operations today. The government could hire him part time, and take advantage of his natural comings and goings in the country where he works. The businessman, artist, or scientist who is a part-time intelligence officer or clandestine operator for the military services is a new idea that is bound to raise every issue imaginable about liability and conflicts of interest. But that's why we have the lawyers in the loop. We can make it work.

Chapter 18

FORWARD FROM HERE

Six months after the September 11 attacks, I ran into a CIA station chief; call her Jenny. It was my first time in New York since the World Trade Center disaster. We happened to be in the city at the same time and started talking about the changes we had seen since the country went to war against Al Qaeda.

"We've seen a change in some attitudes," Jenny said. "It used to be that when we talked to businessmen and asked if they could help us, they would back away. They would say that they had a fiduciary responsibility to their stockholders."

What Jenny was saying was that in the past it had often been hard to get an executive from a U.S. company to help us get access to a target or information. The argument was that, legally, the executive was supposed to look out for the interests of his investors, and working with the CIA or some other clandestine U.S. government agency meant additional risk.

When a company, especially a publicly traded company, looks for investors, it has to disclose any risk factors it is aware of that might hurt the business. Some lawyers might say that if the company has a secret relationship with the government, that's a risk. If the relationship were exposed, the company might get kicked out of a foreign country, or the media might put the company in the spotlight. Either way, it could mean lost business. The company would have trouble making its numbers, the investors could think they were ill served. Which could mean a lawsuit.

In the past, this is the point where a company executive would say, "Why bother?"

"Now," Jenny said, "they're saying, 'Ensuring a stable business environment might be part of our fiduciary responsibilities.'" After the September attacks, the Dow dropped 1,000 points. Consumer confidence went in the tank, air travel was a bitch, and companies had to spend millions to beef up their security—and work around all the security measures everyone else adopted.

Now company executives were thinking that, patriotism aside, helping the government track down terrorists and anyone else that might attack the United States was good for business. So now, if someone asked why they were taking risks to help the government, they had a defense. Some of them couldn't wait to suit up.

It is not just access to communications, although that is critical to victory in network warfare. If anything should be clear by now, network warfare means that there will be no front line in future wars. The enemy can be all around you. And if you hope to win, you need to be able to get all around your enemy. U.S. forces will often need to blend into the background before and after they fight. In many parts of the world, the American private sector has better access than any government agency.

It's really hard to tell if the gung-ho, let-me-at-them spirit will hold up. It started to sag once the Taliban regime fell and the news media turned its attention to Enron and Worldcom. This would be too bad, because government officials and the private sector will need to work more closely than ever if we expect to deal effectively with threats to the United States, especially the threats that we face today. Rogue states like Iraq, Iran, Libya, and North Korea, terrorist organizations like Al Qaeda, and strategic rivals like China all are skilled in concealment and are developing greater expertise in the new forms of warfare.

After the September 11 attacks, many experts said that we needed more cooperation between the federal government and the private sector. They were mainly talking about homeland defense. The kind of cooperation I am talking about here is different.

Homeland defense is border patrols and airport security, or law enforcement and emergency rescue. These are all important. But the government and the private sector also need to work more closely together to win foreign wars.

Time has shown that throughout the world, as people learn of the benefits of democracy, capitalism, and the rule of law, they adopt those institutions and become fat and happy. They vote, raise families, and make money. They hope to succeed, but even if they don't achieve all their dreams, they would rather lose and have another turn at the game of life rather than resort to crime, terror, or war. That is the defining feature of modern society.

When all peoples achieve this goal, we will be able to rely on diplomacy and law enforcement to keep the peace. Until then, we all need to prepare for combat in the Information Age.

ACKNOWLEDGMENTS

Many thanks are due to John Raisian, the director of the Hoover Institution. John was generous in both his support and his encouragement, and I thank him, and all my other Hoover colleagues.

Readers will note that the RAND Corporation occasionally appears in some of the events recounted here. I've been affiliated with RAND, too, for more than a decade. It's impossible to tell a story like this without mentioning RAND. This simply reflects the unique and important role RAND has played in the development of U.S. national security policy during the past half-century. I appreciate all the encouragement and insights all of my colleagues have given me over the years. Needless to say, however, this book is not a RAND publication, and does not represent the views of RAND, its clients, or for that matter, any of the various government organizations I have been associated with.

This book would not have been possible without information, comments, and interviews provided by many people who graciously offered their time, and I would like to thank William Adams, William Ailor, Martin Anderson, Rich Arenberg, John Arquilla, Bruce Artman, Robert Baer, John Baker, Robert Ballard, Richard Best, Greg Blackburn, William Broad, Robert Carpenter, Susan Chema, Richard Cronin, William Crowell, Alan Cullison, Jack David, Douglas Dearth, Arnaud de Borchgrave, Dorothy Denning, David Deptula, James Fallows, Jennette Finley, Dan Freedberg, John Garstka, Frederick Giessler, Christopher Glaze, Robert Hahn, Grant Hammond, John

Hamre, Martin Hill, Harry Hillaker, Ralph Hitchens, Bruce Hoffman, Fred Ickle, Ronald Knecht, Andrew Krepinevich, Trudy Kuehner, Francis LaCroix, Anthony Lazarski, Martin Libicki, Thomas Longstaff, David Lonsdale, Andrew Marshall, Walter McDougall, Harris Miller, Roger Molander, Stephen Morrow, James Mulvenon, William Nolte, James O'Brien, Kevin O'Connell, Catherine Offinger, Richard O'Neill, Emmitt Paige, Joe Pendry, Jeffrey T. Richelson, Ervin Rokke, Michael Rona, Thomas Rona, Jr., David Ronfeldt, Henry Rowen, Kevin Ruffner, Franklin Spinney, James Steinberg, Kenneth Stringer, Bernard Trainor, Michelle van Cleve, William van Cleve, Paul van Riper, Caroline Wagner, Carolyn Warner, Michael Warner, Larry Welch, Robert West, Albert Wheelon, Peter Wilhelm, Peter Wilson, John Woodward III, Bruce Wright, and Philip Zelikow. Annelise Anderson and Christopher Mellon reviewed drafts of the entire manuscript and provided comments, much to my benefit.

As is usual in books of this kind, a handful of people who assisted asked not to be mentioned, and I extend my appreciation to them as well.

Finally, special thanks go to Lee, Theresa, Kelley, and most especially, Bob.

NOTE ON INFORMATION SOURCES

This book is based on official documents, press reports, and interviews. In compliance with security agreements that come with one's working in the defense and intelligence fields, everything is based on sources that are available to the public. Where the book refers to a possibly classified activity, a citation to the appropriate reference is included. The reader should assume that mention in this book is not intended as confirmation of any classified information.

REFERENCES

CHAPTER 1

1. Prices as of late 2002 were as follows: an ATN MO 3-3D night-vision scope cost $1,195; a Magellan model MA 59801 handheld, mapping Global Positioning System receiver cost $295; and a satellite image with .6-meter spatial resolution from DigitalGlobe cost $6,100. As they say, your cost may vary.

CHAPTER 2

1. This account is based on interviews with Alan Cullison. Also see his articles with Andrew Higgins, "Forgotten Computer Reveals Thinking Behind Four Years of Al Qaeda Doings," *Wall Street Journal,* December 31, 2001, and "Account of Spy Trip on Kabul PC Matches Travels of Richard Reid," *Wall Street Journal,* January 16, 2002; Dan Kennedy, "How the *Journal* Got Al Qaeda's Computers," posted on ThePhoenix.com Web site at www.bostonphoenix.com/boston/news_features/this_justin/documents/02093041.htm; Felicity Barringer, "Why Reporters' Discovery Was Shared With Officials," *New York Times,* January 21, 2002;
2. See, for example, James Risen and David Johnston, "Intercepted Al Qaeda E-Mail Is Said to Hint at Regrouping," *New York Times,* March 6, 2002.
3. Information on el-Hage's background, the raid on the Nairobi safe house, and the exploitation of the Nairobi computer is taken from Phil Hirschkorn, "Passport Offers Peek Inside bin Laden's Businesses," *CNN Law,* February 26, 2001; Oriana Zill, "A Portrait of Wadih el-Hage, Accused Terrorist," at www.pbs.org/wgbh/pages/frontline/shows/binladen/upclose/elhage.html; and *Transcript of the Trial of Suspected Al Qaeda Militants in Connection With the Bombings of the American Embassies in Kenya and Tanzania on 7 August 1998,* U.S. embassy bombing transcript, days 8–12, February 21, 2001, available at www.cryptome.org.
4. Interview with Larry C. Johnson, "Hunting bin Laden," *Frontline,* September

12, 2001. Transcript is available at www.pbs.org/wgbh/pages/frontline/shows/
binladen/interviews/newjohnson.html.

5. See testimony of Jamal Ahmed al-Fadl, *United States of America v. Usama bin Laden, et al.,* S(7) 98 Cr. 1023, U.S. District Court, Southern District of New York, February 6, 2001, available at www.cryptome.org. Also see Benjamin Weiser, "Ex-Aide to bin Laden Describes Terror Campaign Aimed at U.S.," *New York Times,* February 7, 2001.

6. See Matthew Brzezinski, "Bust and Boom," *The Washington Post Magazine,* December 30, 2001.

7. Criminal complaint, *United States of America v. Richard C. Reid,* filed in U.S. District Court, District of Massachusetts, December 22, 2001.

8. Josh Meyer and Sebastian Rotella, "Passenger Likely Not Al Qaeda, U.S. Says," *Los Angeles Times,* December 24, 2001.

9. For casualties at Antietam, see the National Park Service Web site at www.nps.gov/anti/casualty.htm. For Pearl Harbor, see Gordon W. Prange, *At Dawn We Slept: The Untold Story of Pearl Harbor,* New York: Viking, 1991. For the September 11 attack, see Wikipedia, the online resource at www.wikipededia.org/wiki/September_11_2001_Terrorist_Attack/Casualties; Wikipedia has updated its estimate as additional information has become available from the government and major media organizations. This estimate was last updated on October 8, 2002. It does not include the nineteen terrorists who died in the attack. Note that, unlike the other attacks, in which almost all casualties were military personnel, the vast majority of casualties in the September 11 strike were civilian, and many were foreigners.

10. For insurance losses resulting from the attack, see, for example, *Implications of the September 11 Terrorist Attacks,* New York: Tillinghast-Towers Perrin Reinsurance, September 21, 2001. For the effects of the attack on the New York City economy, see *Economic Impact Analysis of the September 11th Attack on New York,* New York City Partnership and Chamber of Commerce, November 16, 2001. For costs of repairing the Pentagon, see Gerry J. Gilmore, "Pentagon Terror Damage Will Take Years to Repair," American Forces Press Service, October 3, 2001. For stress reactions to the attack, see Mark A. Schuster et al., "A National Survey of Stress Reactions After the September 11, 2001, Terrorist Attacks," *New England Journal of Medicine,* November 15, 2001, pp. 1507–1512.

11. David Kastenbaum, "Mini Nukes," report on *Weekend All Things Considered,* April 7, 2002. Also see Bill Gertz, "Moscow Builds Bunkers Against Nuclear Attack," *Washington Times,* April 1, 1997, which cites a CIA report on recent Russian efforts, and Stephen I. Schwartz, "This Is Not a Test," *Bulletin of the Atomic Scientists,* November/December 2001, on recent U.S. activities.

12. For press reports of SIGINT intercepts of Al Qaeda communications and Taliban preparation, see Rowan Scarborough, "Intercepts Foretold of 'Big Attack,'" *Washington Times,* September 22, 2001. Note that these reports occurred just a week after the Al Qaeda attacks and the source of the infor-

mation was described as a "senior administration official."In June 2002, Bush administration officials strongly criticized congressional committees investigating the failure of U.S. intelligence to anticipate the attacks for leaking this information. In fact, the original leak appears to have come from the administration itself, at a time when Al Qaeda's role was not clear and the administration was seeking to justify the case for taking military action. The later leaks, while offering slightly more detail, were mainly a rehash. Binalshibh told an al-Jazeera interviewer of Atta's message on the day of the attack; see Susan Schmidt and Dan Eggen, "Suspected Planner of 9/11 Attacks Captured in Pakistan After Gunfight," *The Washington Post,* September 14, 2002.

13. For the final preparations of the terrorists, see Don Van Natta, Jr., and Kate Zernike, "Hijackers' Meticulous Strategy of Brains, Muscle and Practice," *New York Times,* November 4, 2001.

CHAPTER 3

1. See R. V. Jones, *The Wizard War: British Scientific Intelligence 1939–1945,* New York: Coward, McCann & Geoghegan, 1978, pp. 84–87, 135–150, and 161–178.
2. German losses are based on German records and cited in W. F. Craven and J. L. Cates, eds., *The Army Air Forces in World War II: Plans and Early Operations, January 1939 to August 1942,* Washington: U.S. Government Printing Office, 1951, p. 95.
3. John McCarthy, "Memorandum from John McCarthy to P. M. Morse: Subject: A Time-Sharing Operator Program for Our Projected IBM 709," January 1, 1959. Morse was the director of the MIT Computation Center, which was scheduled to receive the new computer. The memorandum is reproduced on McCarthy's home page. See www-formal.stanford.edu/jmc/history/timesharing-memo.html.
4. This is taken from David Allison's interview, *Transcript of a Video History Interview with Mr. William "Bill" Gates,* Washington, D.C.: Division of Computers, Information, & Society, National Museum of American History, Smithsonian Institution, available at americanhistory.si.edu/csr/comphist/gates.htm#tc46. Also see James Wallace, *Hard Drive: Bill Gates and the Making of the Microsoft Empire,* New York: HarperCollins, 1993.
5. Thomas P. Rona, *Weapons Systems and Information War,* Seattle: Boeing Aerospace Company, July 1976, pp. 1–2.

CHAPTER 4

1. See A. W. Marshall, "Long Term Competition with the Soviet Union: A Framework for Strategic Analysis," Santa Monica, CA: RAND, April 15, 1971. Declassified July 5, 1980. This was an unpublished internal note; the contents cited here were described in interviews.
2. See David R. Beachley, "Extending the Battlefield: Soviet Radio-Electronic Combat in World War II," *Military Review,* March 1981.
3. Letter from Thomas P. Rona to Andrew Marshall, July 26, 1976.

CHAPTER 5

1. For a comparison of the two aircraft, see "MiG-17 vs. F-105," prepared by the U.S. Air Force Museum and available at www.wpafb.af.mil/museum/ history/ Vietnam/469th/p39.htm.
2. Harry Hillaker in "Tribute to John R. Boyd," *Code One Magazine,* July 1997, and interview.
3. See Grant T. Hammond, *The Mind of War: John Boyd and American Security,* Washington, D.C.: Smithsonian Institution Press, 2001. Hammond is a professor at the Air War College and was a personal acquaintance of Boyd's. Jeffrey L. Cowan has written a master's thesis that provides a concise biography and an account of Boyd's role in developing the current Marine Corps; see his *From Air Force Fighter Pilot to Marine Corps Warfighting: Colonel John Boyd, His Theories on War, and Their Unexpected Legacy,* U.S. Marine Corps Command and Staff College (academic year 1999–2000), available at www.d-n-i.net/fcs/ boyd_thesis.htm#ex sum. Also see James G. Burton, *The Pentagon Wars: Reformers Challenge the Old Guard,* Annapolis, MD: Naval Institute Press, 1993, and the forthcoming biography by Robert Coram: *Boyd: The Fighter Pilot Who Changed the Art of War,* Boston: Little, Brown, 2002. Excerpts are available at www.robertcoram.com.
4. Grant T. Hammond interview, February 27, 2002; and Franklin C. Spinney interview, March 1, 2002.
5. See John R. Boyd, "A Discourse on Winning and Losing," unpublished manuscript, August 1987, provided by Grant T. Hammond.
6. Spinney and Hammond interviews. Boyd influenced, and was influenced by, academic studies of decision making and, in particular, the area of study known as bounded rationality. This field is usually linked with Herbert Simon, who proposed it as a supplement to the traditional model of the rational, utility-maximizing actor. According to this model, otherwise rational actors are led to other-than-optimal choices because they have previously (or implicitly) agreed to lock themselves into a decision-making process for reasons of efficiency (which, in itself, is a rational choice). See Herbert A. Simon, *Administrative Behavior,* 2nd ed., New York: Macmillan, 1957. For some applications and analyses in military situations, see, for example, the following two theses by students at the School of Advanced Airpower Studies, Air University, Maxwell Air Force Base, Alabama: Arden B. Dahl, "Command Dysfunction: Minding the Cognitive War," June 1996; and Michael T. Plehn, "Control Warfare: Inside the OODA Loop," June 2000.

CHAPTER 6

1. As has been noted previously, all references to CIA activities can be sourced to publicly available reports. In this case, see Douglas Waller, "Inside the CIA's Covert Forces," *Time,* December 10, 2001; Dana Priest, "Team 555 Shaped a

New Way of War," *The Washington Post,* April 3, 2002; and Judith Miller and Eric Schmitt, "Ugly Duckling Turns Out to Be Formidable in the Air," *New York Times,* November 23, 2001. Note that Miller and Schmitt attribute their information about the CIA's advocating arming the Predator to intelligence officials.

2. See Commission on Roles and Capabilities of the United States Intelligence Community, *Preparing for the 21st Century: An Appraisal of U.S. Intelligence,* Washington, D.C.: U.S. Government Printing Office, 1996, p. 132. The graph on that page suggests that the CIA budget is around $3.1 billion. Although that amount has changed, the comparison with the Defense Department budget remains valid, if only because of the relative size of both. Also see R. Jeffrey Smith, "Making Connections With the Dots to Decipher U.S. Spy Spending," *The Washington Post,* March 12, 1996.

3. Floor speech by Porter Goss, *Congressional Record,* December 11, 2001, pp. H9149–H9152.

4. Edwin C. Fishel, *The Secret War for the Union: The Untold Story of Military Intelligence in the Civil War,* Boston: Houghton Mifflin, 1996, pp. 347–349.

5. Gregory R. Blackurn interview, July 17, 2001.

6. See Jeffrey T. Richelson, *The U.S. Intelligence Community,* New York: Ballinger, 1989, p. 197, and Bob Woodward, *Veil: The Secret Wars of the CIA 1981–1987,* New York: Simon & Schuster, 1987, pp. 448–449.

7. Transcript of *Nova* episode 2602, "Submarines, Secrets, and Spies." Air date, Jan. 19, 1999. Codevilla also discussed the project in *Informing Statecraft: Intelligence for a New Century,* New York: The Free Press, 1992. Bob Woodward describes the role of John Butts and E. A. Burkholder in the project in *Veil: The Secret Wars of the CIA 1981–1987,* New York: Simon & Schuster, 1987, pp. 449–450. Also see Sherry Sontag and Christopher Drew, with Annette Lawrence Drew, *Blind Man's Bluff: The Untold Story of American Submarine Espionage,* New York: Public Affairs, 1998, pp. 247–248; and William J. Broad, *The Universe Below: Discovering the Secrets of the Deep Sea,* New York: Simon & Schuster, 1997, pp. 81–84.

8. Most of what is publicly known about U.S. communications interception activities has been the result of espionage cases and press leaks, combined with information about "national technical means" that has been released to facilitate arms control verification. An overview of capabilities compiled from these and other sources (useful for illustration, although no one with actual access can vouch for its completeness or accuracy) appears in Jeffrey T. Richelson, *The U.S. Intelligence Community,* 4th ed., Boulder, CO: Westview Press, 1999.

9. Ronald Knecht interview, September 21, 2001.

10. The JSSG is cited in Ronald Knecht and Ronald A. Gove, *The Information Warfare Challenges of a National Information Infrastructure,* available at www.infowar.com/mil_c4i/informationwarfarechall.html-ss. Also see the Air Force biography of Brigadier General Grover E. Jackson, who served as vice director of the JSSG from June 1987 to June 1989, available at www.af.mil/news/biographies/jackson_ge.html.

11. David A. Fulghum and Robert Wall, "Information Warfare Isn't What You Think," *Aviation Week & Space Technology*, February 26, 2001.

12. Special technical operations and the Planning and Decision Aid System are cited in *Joint Doctrine for Command and Control Warfare (C2W)*, Joint Pub. 3–13, Washington, D.C.: Department of Defense, Joint Chiefs of Staff, February 7, 1996.

CHAPTER 7

1. See the Free Congress Foundation Web site at www.freecongress.org/fcf/

2. This quotation is from Peter Golden, "The Partisan," *Electronic Business*, August 1999. The rest of the material is from an interview with Strassmann on September 11, 2001.

3. Duane Andrews interview, February 11, 2002.

CHAPTER 8

1. David Ronfeldt, "Social Science at 190 M.P.H. on NASCAR's Biggest Superspeedways," *First Monday*, February 2000, available at firstmonday.org/issues/issue5_2/ronfeldt/index.html.

2. See Bob Woodward, *The Commanders*, New York: Pocket Books, 1992, p. 219.

3. There are several accounts of how the Left Hook was conceived. This is likely an example of "victory finds a hundred fathers, while defeat is an orphan." Michael Gordon and Bernard Trainor, who do not have a dog in the fight, offer their analysis in *The Generals' War: The Inside Story of the Conflict in the Gulf*, Boston: Little, Brown, 1995. They give special emphasis to Rowen's contribution. For the official Army history, see Robert H. Scales, *Certain Victory: The U.S. Army in the Gulf War*, Washington: Brassey's, 1994. For Schwarzkopf's account, see H. Norman Schwarzkopf, *It Doesn't Take a Hero*, New York: Bantam, 1992. Also see *Frontline* interview with Norman Schwarzkopf, available at www.pbs.org/wgbh/pages/frontline/gulf/oral/schwarzkopf/1.html

4. See the battalion history of the 1-101st Aviation Regiment, Fort Campbell, KY, available at www.campbell.army.mil/1101avn/history.htm. Also see the account provided by Richard Mackenzie, "Apache Attack," *Air Force Magazine*, October 1991.

5. See H. Norman Schwarzkopf, *It Doesn't Take a Hero*, pp. 522–547.

6. See Larry K. Wentz, "Communications Support for the High Technology Battlefield," in Alan D. Campen, ed., *The First Information War: The Story of Communications, Computers and Intelligence Systems in the Persian Gulf War*, Fairfax, VA: AFCEA International Press, 1992.

7. See interview of Frederick T. Andrews, Jr., by David Hochfelder, New Brunswick, NJ: IEEE History Center, Rutgers University, 1999. Andrews was a top electrical engineer and executive at Bell Laboratories.

8. Paul Baran, *On Distributed Communications: I. Introduction to Distributed*

Communications Network. Memorandum RM-3420-PR, Santa Monica, CA: RAND, August 1964.

9. See Bruce Sterling, "Short History of the Internet," *The Magazine of Fantasy and Science Fiction,* February 1993, available at nethistory.html; and Leonard Kleinrock, "The Day the Infant Internet Uttered Its First Words," available at www.lk.cs.ucla.edu/LK/Inet/1stmesg.html.

10. See John Arquilla and David Ronfeldt, "Cyberwar Is Coming!" *Comparative Strategy,* Spring 1993, pp. 141–165. Arquilla and Ronfeldt expanded on the idea in a later collection of essays, *Networks and Netwars: The Future of Terror, Crime, and Militancy,* Santa Monica, CA: RAND, 2001, available online at www.rand.org/publications/MR/MR1382/. Also John Arquilla and David Ronfeldt interviews.

CHAPTER 9

1. See Martin L. Van Creveld, *Technology and War,* New York: Macmillan, 1989; and Andrew Vick, "Medieval Siege Technology and Countertechnology," at web.grinnell.edu/techstudies/vick/index.htm.

2. See Richard Bak, *The C.S.S. Hunley: The Greatest Undersea Adventure of the Civil War,* Dallas: Taylor Publishing, 1999. The *Hunley* was located in May 1995 and raised in August 2000. At this writing, marine archeologists are analyzing the wreck to assess how the submarine was lost. For a summary, see Glenn Oeland, "The H. L. Hunley: Secret Weapon of the Confederacy," *National Geographic,* July 2002, pp. 82–101, and the Web site of the recovery effort, www.hunley.org.

3. *Report of Lieutenant W. B. Cushing, Albemarle Sound, N.C.* Recorded by Asst. Engineer Ezra Jabex Whitaker of the U.S.S. *Lackawana,* October 30, 1864. Provided by his granddaughter, Sara Whitaker Hale, and posted at www.brownwaternavy.com.

4. The definitive source on Robert Whitehead (on which much of this section is based) is Edwyn Gray, *The Devil's Device: Robert Whitehead and the History of the Torpedo,* Annapolis, MD: Naval Institute Press, 1975.

5. Geoff Kirby, "A History of the Torpedo: The Early Days," *Journal of the Royal Navy Scientific Service* (Vol. 27, No. 1) at www.btinternet.com/-philipr/torps.htm.

6. See L. S. Howeth, *History of Communications-Electronics in the United States Navy,* Washington, D.C.: Bureau of Ships and Office of Naval History, 1963, Chapter 24. Inevitably, the army had its own version of the idea, called the Bug, which it tested at about the same time.

7. James E. Tomayko, "Helmut Hoelzer's Fully Electronic Analog Computer," *Annals of the History of Computing,* July 1985, pp. 227–240.

8. Kenneth P. Werrell, "Did USAF Technology Fail in Vietnam? Three Case Studies," *Airpower Journal,* Spring 1998.

CHAPTER 10

1. *Frontline* interview with H. Norman Schwarzkopf, available at www.pbs.org/wgbh/pages/frontline/gulf/oral/schwarzkopf/5.html.

2. Buster C. Glosson, "Impact of Precision Weapons on Air Combat Operations," *Airpower Journal,* Summer 1993. The article was based on a presentation Glosson made at the 1992 Armament Symposium at Eglin Air Force Base, September 23, 1992.

3. The memo is reproduced as Exhibit III in Cynthia Ingols and Lisa Brem, *Implementing Acquisition Reform: A Case Study on Joint Direct Attack Munitions,* Fort Belvoir, VA, Defense Systems College Management, 1998.

4. J. J. O'Connor and E. F. Robertson provide a very thorough discussion of the development of longitude and measurement of the earth in the *MacTutor History of Mathematics,* maintained by the School of Mathematical and Computational Sciences at the University of St. Andrews. It is available at www-history.mcs.st-and.ac.uk/history/. The classic reference for the history of maps and cartography is Lloyd A. Brown, *The Story of Maps,* New York: Dover, 1977, an update of his 1949 work.

5. Benjamin Franklin (Margaret K. Soifer, ed.), *The Autobiography of Benjamin Franklin,* New York: Macmillan, 1967.

6. See the Web site of H.M. Nautical Almanac Office at www.nao.rl.ac.uk/nao/history/.

7. See Ron Cardoza, *Evolution of the Sextant,* San Diego, CA: West Sea Company, 2000; and J. J. O'Connor and E. F. Robertson, "Longitude and the Académie Royale," and "English Attacks on the Longitude Problem," in the *MacTutor History of Mathematics, op. cit.*

8. Dava Sobel wrote a best-selling popular account of Harrison's travails; see her *Longitude: The True Story of a Lone Genius Who Solved the Greatest Scientific Problem of His Time,* New York: Walker & Co., 1995.

9. See Bruce D. Berkowitz, *American Security: Dilemmas for a Modern Democracy,* New Haven, CT: Yale University Press, 1986, Chapter 5; and Pat Hillier and Nora Slatkin, *U.S. Ground Forces: Design and Cost Alternatives for NATO and Non-NATO Contingencies,* Washington, D.C.: Congressional Budget Office, 1980.

10. See Stephen P. Rosen, *Winning the Next War: Innovation and the Modern Military,* Ithaca, NY: Cornell University Press, 1991; and Barry D. Watts, "What Is the 'Revolution in Military Affairs'?" Northrop Grumman Analysis Center monograph, April 5, 1995.

11. See U.S. General Accounting Office, *Weapons Acquisitions: Guided Weapon Plans Need to Be Reassessed,* GAO/NSIAD 99-32, December 1998; and Eric Schmitt, "Improved U.S. Accuracy Claimed in Afghan Air War," *New York Times,* April 9, 2002.

CHAPTER 11

1. See U.S. Department of Defense, "Navy Announces Results of Its Investigation on *U.S.S. Cole,* press release, January 19, 2001; also see *DOD U.S.S. Cole Commission Report,* Washington, D.C.: Department of Defense, January 9, 2001; and Steven Lee Myers, "After Cole's Bombing, Pentagon Finds Ongoing Lapses in Gulf Security," *New York Times,* January 1, 2001.

2. See John Arquilla, David Ronfeldt, *Swarming and the Future of Conflict,* Report DB-311-OSD, Santa Monica, CA: RAND, 2000. Also see Sean J. A. Edwards, *Swarming on the Battlefield: Past, Present, and Future,* Report MR-1100-OSD, Santa Monica, CA: RAND, 2000.

3. Max Boot, *The Savage Wars of Peace: Small Wars and the Rise of American Power,* New York: Basic Books, 2002.

4. U.S. Marine Corps, *Small Wars Manual,* Washington, D.C.: Government Printing Office, 1940. The manual was reissued by the Marines as NAVMC 2890 in 1987 and distributed commercially by Sunflower University Press in 1989.

5. The Boston Study Group, *Winding Down: The Price of Defense,* New York: Times Books, 1979.

6. U.S. Marine Corps, *Warfighting* (MCDP1), Washington, D.C.: Department of the Navy, Headquarters, U.S. Marine Corps, first edition 1989. A second edition was issued in 1997 under Gray's successor, Charles Krulak; this is essentially a refinement and an update, and was countersigned by Gray.

7. U.S. Marine Corps, *Command and Control* (MCDP6), Washington, D.C.: Department of the Navy, Headquarters, U.S. Marine Corps, 1996. The Marines issued Marine Corps Doctrinal Publication 6 in 1996 under Krulak; it includes explicit references to Boyd and the OODA loop.

8. Thomas E. Ricks and Bob Woodward, "Marines Enter South Afghanistan," *The Washington Post,* November 26, 2001. An Army Ranger team raided an Al Qaeda base in southern Afghanistan on the night of October 19, but this was a short hit-and-run operation.

9. For a pithy analysis of the whole sweatshop phenomenon, see Nicholas D. Kristof, "Let Them Sweat," *New York Times,* June 25, 2002.

10. See Daniel A. Wren and Ronald G. Greenwood, *Management Innovators: The People and Ideas That Have Shaped Modern Business,* New York: Oxford University Press, 1998; and Ronald M. Becker, "Lean Manufacturing and the Toyota Production System," SAE International, available at www.sae.org/topics/leanjun01.htm.

11. James P. Womack, Daniel T. Jones, and D. Roos, *The Machine That Changed the World: The Story of Lean Production,* New York: HarperCollins, 1991; and Taiichi Ohno, *Toyota Production System: Beyond Large-Scale Production,* Cambridge, MA: Productivity Press, 1988.

12. John Garstka interview, March 26, 2002.

13. Arthur K. Cebrowski and John J. Garstka, "Network-Centric Warfare: Its Origin and Future," *Proceedings of the U.S. Naval Institute,* January 1998. Also see David S. Alberts, John J. Garstka, Richard E. Hayes, and David A. Signori, *Un-*

derstanding Information Age Warfare, Washington, D.C.: Department of Defense, Command and Control Research Program, 2001.

14. Peter Steiner, *The New Yorker,* July 5, 1993, p. 61.
15. See, for example, Dana Priest, "'Team 555' Shaped a New Way of War: Special Forces and Smart Bombs Turned Tide and Routed Taliban," *The Washington Post,* April 3, 2002; David Rohde, "Anatomy of a Raid in the Afghan Mountains," *New York Times,* June 3, 2002.
16. Quotations from people who took part in the battle are taken from interviews that appeared in the documentary *The True Story of Black Hawk Down,* produced by A&E Films, 2000. See also Mark Bowden, *Black Hawk Down: A Story of Modern War,* Boston: Atlantic Monthly Press, 1999. Bowden participated in the production of the A&E documentary.

CHAPTER 12

1. Judith Miller and Eric Schmitt, "Ugly Duckling Turns Out to Be Formidable in the Air," *New York Times,* November 23, 2001.
2. See Seymour M. Hersh, "Annals of National Security: King's Ransom: How Vulnerable Are the Saudi Royals?" *The New Yorker,* October 22, 2001; and Judith Miller and Eric Schmitt, "Ugly Duckling Turns Out to Be Formidable in the Air," *New York Times,* November 23, 2001.
3. Bob Woodward, "Lethal Covert Action," *The Washington Post National Weekly Edition,* October 29–November 4, 2001.
4. Jeffrey T. Richelson has written an especially thorough review of assassination and assassination policies in recent history; see his "When Kindness Fails: Assassination as a National Security Option," *International Journal of Intelligence and CounterIntelligence,* Winter 2002, pp. 243–274.
5. See Bernard Lewis, *The Assassins—A Radical Sect in Islam,* London: Oxford University Press, 1967; also see the etymology provided in the *American Heritage Dictionary of the English Language,* 4th ed., New York: American Heritage, 2000.
6. See Bruce D. Berkowitz and Allan E. Goodman, "The Logic of Covert Action," *The National Interest,* Spring 1998, pp. 38–46; and Barton Gellman, "Clinton's War on Terror," *The Washington Post,* December 20, 2001.
7. See U.S. Congress, Senate Select Committee to Study Governmental Operations with Respect to Intelligence Activities, *Interim Report: Alleged Assassination Plots Involving Foreign Leaders,* 94th Cong., 1st Sess., S. Report No. 94-465, Washington, D.C.: GPO, 1975; Gregory F. Treverton, *Covert Action: The Limits of Intervention in the Postwar World,* New York: Basic Books, 1987; and William E. Colby, with Peter Forbath, *Honorable Men: My Life in the CIA,* New York: Simon & Schuster, 1978.
8. See the account in William Stevenson, *A Man Called Intrepid: The Secret War,* Guilford, CT.: The Lyons Press, 1976.
9. See William A. Tidwell, with James O. Hall and David Winfred Gaddy, *Come*

Retribution: The Confederate Secret Service and the Assassination of Lincoln, Oxford, MS.: University Press of Mississippi, 1988.

10. Transcript of President Bush's address to a joint session of Congress, September 20, 2001, from *The Washington Post,* September 21, 2001.

11. "Bush's United States Military Academy Graduation Speech," transcript in *The Washington Post,* June 2, 2002.

12. Eric Schmitt, "U.S.-Philippine Command May Signal War's Next Phase," *New York Times,* January 16, 2002; Bob Woodward, "President Broadens Anti-Hussein Order; CIA Gets More Tools to Oust Iraqi Leader," *The Washington Post,* June 16, 2002.

CHAPTER 13

1. Charles Piller, "Power Grid Vulnerable to Hackers," *Los Angeles Times,* August 23, 2001.

2. For an example of a typical Day After . . . game and accompanying analysis of the results, see Roger Molander, Andrew S. Riddile, and Peter A. Wilson, *Strategic Information Warfare: A New Face of War,* Report MR-661-OSD, Santa Monica, CA.: RAND, 1996.

3. William Crowell interview, July 19, 2002.

4. James Adams, "Virtual Defense," *Foreign Affairs,* May/June 2001, pp. 98–112.

5. The comment to the telecommunications executives was cited by, among others, Eun-Kyung Kim. "Cyber Threats: Internet Addiction Makes America Vulnerable to Cyber Attacks," Associated Press report, November 4, 1999. Clarke made the "tsunami" statement at a meeting of the American Bar Association in July 1998; reported by David Ruppe, "Cyber Scare: Some Experts Say Government Officials Overstate Computer Threats," ABCNEWS.com, August 4, 1999; available at abcnews.go.com/sections/world/DailyNews/cyber990804.html.

6. See Rob Rosenberger, "When Did the Term 'Computer Virus' Arise?" *Scientific American,* online edition, September 2, 1997, available at www.sciam.com/askexpert/computers/computers15/. Also see the vMyths.com home page, which Rosenberger maintains with George Smith.

7. Michelle Delio, "The Greatest Hacks of All Time," *Wired,* February 6, 2001, available at www.wired.com/news/technology/0,1282,41630,00.html.

8. Jeri Clausing, "In Hearing on 'Love Bug,' Lawmakers Go After Software Industry," *New York Times,* May 11, 2000.

9. P. J. Hebert, J. D. Jarrell, and M. Mayfield, NOAA Technical Memorandum NWS NHC-31, National Oceanic and Atmospheric Administration, Washington, D.C.: U.S. Department of Commerce, 1992.

10. See *Implications of the September 11 Terrorist Attacks,* New York: Tillinghast-Towers Perrin Reinsurance, September 21, 2001.

11. See Jeff Wynne, "White House Advisor Richard Clarke Briefs Senate Panel on Cybersecurity," *Washington File,* Washington, D.C.: U.S. Department of State International Information Programs, February 14, 2002, available at usinfostate.

gov; and press release, Office of Senator Charles Schumer, "Senate Hearing Reveals New Evidence Showing U.S. Vulnerable to Massive Cyber-Terrorist Attack From Iran, Iraq, Al Qaeda," February 13, 2002.

12. James Steinberg interview, March 7, 2002.

13. William Arkin and Robert Windrem, "The Other Kosovo War," MSNBC transcript, August 29, 2001.

14. Arkin and Windrem, *op cit.*

15. Douglas Waller, "Tearing Down Milosevic; Washington Resorts to a Bag of Tricks to Try to Get Yugoslavia a New Leader," *Time,* July 12, 1999. Also see Philip Sherwell, Julius Strauss and Sasa Nikolic, "Clinton Orders 'Cyber-Sabotage' to Oust Serb Leader," *Daily Telegraph,* July 4, 1999.

16. Quoted in Bradley Graham, "Military Grappling With Guidelines for Cyber Warfare," *The Washington Post,* November 8, 1999.

17. For an analysis of the NATO air campaign, see Benjamin S. Lambeth, *NATO's Air War for Kosovo: A Strategic and Operational Assessment,* Report MR-1365-AF, Santa Monica, CA: RAND, 2001.

18. James O. Ellis, Jr., "A View From the Top," (undated briefing, received December 16, 1999). Ellis was Commander in Chief, U.S. Naval Forces, Europe, and Commander in Chief, Allied Forces, Southern Europe.

19. William A. Arkin, "A Mouse That Roars," special to washingtonpost.com, June 7, 1999.

20. Group interview with Bruce Wright, David Deptula, Christopher Glaze, Anthony Lazarski, Jim Henderson, and other members of Air Combat Command staff, March 8, 2002.

21. Terrie M. Gent, "The Role of Judge Advocates in a Joint Air Operations Center: A Counterpoint of Doctrine, Strategy, and Law," *Aerospace Power Journal,* Spring 1999; and Lee E. Deremer, "Leadership Between a Rock and a Hard Place," *Aerospace Power Journal,* Fall 1996. Interview with General John D. Ryan, May 15–17, 1979, Office of Air Force History, available at columbiad.com/vietnam/articles/ 1997/12972_side.htm.

22. *Aviation Week & Space Technology,* September 15, 1997, pp. 67–68.

23. Office of General Counsel, *An Assessment of International Legal Issues in Information Operations,* Washington, D.C.: Department of Defense, May 1999.

CHAPTER 14

1. See Hector C. Bywater, *The Great Pacific War: A History of the American-Japanese Campaign of 1931–33,* Boston: Houghton Mifflin, 1925; Sir John Hackett et al., *The Third World War: August 1985,* New York: Macmillan, 1978.

2. See President's Critical Infrastructure Protection Board, *The National Strategy to Secure Cyberspace: Draft,* Washington, D.C.: The White House, September 2002. For another example, see John Arquilla, "The Great Cyberwar of 2002," *Wired,* February 1998.

3. A good guide to the basics of hacking is Stuart McClure, Joel Scambray, and

George Kurtz, *Hacking Exposed: Network Security Secrets and Solutions,* Berkeley, CA.: Osbourne/McGraw Hill, 1999. Also see Stephen Northcutt, Donald McLachlan, and Judy Novak, *Network Intrusion Detection: An Analyst's Handbook,* 2nd ed., Indianapolis: New Riders, 2000; Dorothy E. Denning, *Information Warfare and Security,* New York: ACM Press, 1999; and David H. Freedman, "How to Hack a Bank," *Forbes ASAP,* April 3, 2000, available at www.forbes.com/asap/2000/0403/056.html.

4. See Muffett's home page at www.users.dircon.co.uk/~crypto/. Crack—the software, not the drug—is available at www.users.dircon.co.uk/ ~crypto/download/c50-faq.html.

5. For background on SATAN, see www.cerias.purdue.edu/homes/spaf/CS690E/mail/msg00018.html.

6. David K. Every, "The History and Basics of HTML," *MacWeek,* March 9, 2000.

7. Charles Jacobs, "Trusted Operating Systems," SANS Institute Technical Paper, May 14, 2001, available at rr.sans.org/securitybasics/trusted_OS.php.

8. See CNN Report, "Pager Messages Lost in Space," available at www.cnn.com/TECH/space/9805/20/satellite.outage/.

9. Satellite Toolkit is manufactured by Analytical Graphics, Inc. See their Web site at www.stk.com/.

CHAPTER 15

1. For examples of how to read a system log to analyze a hack, see Stephen Northcutt, Donald McLachlan, and Judy Novak, *Network Intrusion Detection: An Analyst's Handbook,* 2nd ed., Indianapolis: New Riders, 2000.

2. See *Defending America's Cyberspace: National Plan for Information Systems Protection, Version 1.0; An Invitation to a Dialogue,* Washington, D.C.: The White House, 2000; and President's Critical Infrastructure Protection Board, *The National Strategy to Secure Cyberspace: For Comment,* Washington, D.C.: The White House, September 2002.

3. See, for example, Charles C. Mann, "Why Software Is So Bad," *Technology Review,* July/August 2002.

CHAPTER 16

1. See Ian Cameron, *To the Farthest Ends of the Earth: The History of the Royal Geographical Society 1830–1980,* London: E. P. Dutton, 1980.

2. Halford J. Mackinder, "The Scope and Methods of Geography" and "The Geographical Pivot of History," reprinted in *Democratic Ideas and Reality,* Westport, CT.: Greenwood, 1981.

   For a good summary of traditional geopolitical theory and an analysis of how it relates to warfare in the Information Age, see David J. Lonsdale, "Information Power: Strategy, Geopolitics and the Fifth Dimension," *Journal of Strategic Studies,* 22.2&3.

3. For a snapshot of Mahan and a concise summary of his career and thinking, see Philip A. Crowl, "Alfred Thayer Mahan: The Naval Historian," in Peter Paret, ed., *Makers of Modern Strategy from Machiavelli to the Nuclear Age*, Princeton, N.J.: Princeton University Press, 1986.

4. Alfred Thayer Mahan, *The Influence of Sea Power Upon History*, Boston: Little, Brown, 1890.

5. Alfred Thayer Mahan, "Subordination in Historical Treatment," President's address to the American Historical Association, December 26, 1902, published in the *Annual Report of the American Historical Association*, 1902.

6. Translated text of the Zimmermann Telegram is from National Archives and Records Administration, General Records of the Department of State, Record Group 59. It is available at www.archives.gov/digital_classroom/lessons/zimmermann_telegram/zimmermann_telegram.html.

7. In fact, the trail leading to the exposure of the message was more complex than this summary might suggest. The Germans also sent a copy, also coded, over commercial lines that went through the United States. The British, not wanting to reveal the true source of their intelligence (and the fact they were intercepting Swedish diplomatic communications), referred the Americans to the copy that was already in the United States. The Americans retrieved the message, matched it to the one the British had, and then confronted the German ambassador in Washington without revealing how they came upon the telegram. At that point, Zimmermann, possibly thinking the message had been leaked from any of a number of possible sources—for example, from someone in the Mexican government—admitted to the plan. See David Kahn, *The Codebreakers: The Comprehensive History of Secret Communications from Ancient Times to the Internet*, rev. ed., New York: Scribner, 1996; and Barbara W. Tuchman, *The Zimmermann Telegram*, New York: Viking, 1958. Also see William F. Friedman and Charles J. Mendelsohn, *The Zimmermann Telegram of January 16, 1917, and Its Cryptographic Background*, Laguna Hills, CA: Aegean Park Press, 1976; and Jeffrey T. Richelson, *A Century of Spies: Intelligence in the Twentieth Century*, New York: Oxford University Press, 1995.

8. Peter Earley interview of Boris Aleksandrovich Solomatin. Available at www.crimelibrary.com/terrorists_spies/spies/solomatin/2.htm. Earley, a reporter with *The Washington Post*, covered espionage and developed contacts with former Soviet intelligence officials. Also see his book, *Confessions of a Spy: The Real Story of Aldrich Ames*, New York: G. P. Putnam's Sons, 1997. Interestingly, Solomatin believed the intelligence Walker provided was much more significant than that provided by Ames (and, presumably, did much more damage to the United States).

9. For an account of the initial relationship between the CIA and Oracle, see Mark Leibovich, "The Outsider, His Business and His Billions," *The Washington Post*, October 30, 2000.

10. See Carl Posey, "Air War in the Falklands," *Air & Space/Smithsonian*, September 2002.

CHAPTER 17

1. See Andrew Higgins and Alan Cullison, "Saga of Dr. Zawahri Sheds Light on the Roots of Al Qaeda Terror," *Wall Street Journal,* July 2, 2002. Quotations are taken from that article. Also see Cullison and Higgans, "One Acolyte of Many Vow to Die for al Qaeda Cause," *Wall Street Journal,* December 30, 2002.

2. Briefing by Rep. Saxby Chambliss (July 17, 2002); cited by Steven Aftergood, *Secrecy News* electronic newsletter, Federation of American Scientists; available in archive at www.fas.org/sgp/news/secrecy/2002/07/071902.html.

3. Transcript from *The O'Reilly Factor,* September 20, 2001. For an account of how the Turner cuts occurred that is more detailed and accurate than usually provided, see Bobby R. Inman, "Spying for a Long, Hot War," *New York Times,* October 9, 2001.

4. Anthony S. Harrington, Lew Allen, Jr., Ann Z. Caracristi, and Harold W. Pote, *Report on the Guatemala Review,* Washington, D.C.: U.S. Intelligence Oversight Board, June 28, 1996; R. Jeffrey Smith, "Two-Year 'Scrub' Finds Poor Sources, Serious Criminals," *The Washington Post,* March 2, 1997; and Tim Weiner, "CIA Breaks Links to 100 Foreign Agents," *New York Times,* March 3, 1997.

5. See Joseph E. Persico, *Piercing the Reich: The Penetration of Nazi Germany by American Secret Agents During World War II,* New York: Viking, 1979; and William Casey, *The Secret War Against Hitler,* New York: Berkley Publishing Group, 1989.

6. See Reuel Marc Gerecht, "Notes & Dispatches Peshawar: The Counterterrorist Myth," *Atlantic Monthly,* July/August 2001, and "Blundering Through History With the CIA," *New York Times,* April 23, 2000. See also his earlier article under the pen name Edward G. Shirley, "Can't Anybody Here Play This Game?" *Atlantic Monthly,* February 1998.

7. E. F. Codd, "A Relational Model of Data for Large Shared Data Banks," *Communications of the ACM,* June 1970.

8. Mark Leibovich, "The Outsider, His Business and His Billions," *The Washington Post,* October 30, 2000. Also see Daniel Morrow, *Oral History Interview With Lawrence Ellison, President and CEO Oracle Corporation.* Interview conducted for the Computerworld Smithsonian Awards Program, October 24, 1995, available at americanhistory.si.edu/csr/comphist/le1.html.

9. See, for example, Andrew Higgins and Christopher Cooper, "CIA-Backed Team Used Brutal Means to Break Up Terrorist Cell in Albania," *Wall Street Journal,* November 21, 2001; and Bob Woodward, "50 Countries Detain 360 Suspects at CIA's Behest," *The Washington Post,* November 22, 2001.

10. For example, see the Web site maintained by the Aerospace Corporation's Center for Orbital and Reentry Debris Studies at www.reentrynews.com/recovered.html. Also see A.L. Moore,. and J.V. Leaphart, "Catch That Falling Star! State Responsibility and the Media in the Demise of Space Objects," *Proceedings of the 26th Colloquium on the Law of Outer Space,* Washington, D.C.:

American Institute for Aeronautics and Astronautics, 1984; and Keith Hall, "Remarks to the National Network of Electro-Optical Manufacturing Technologies Conference, February 9, 1998. Hall was Director of the National Reconnaissance Office at the time; the NRO had become the parent organization of the NRL division that developed GRAB.

11. See Naval Research Laboratory, Naval Center for Space Technology, *GRAB: Galactic Radiation and Background; World's First Reconnaissance Satellite* information sheet, available at www.ncst.nrl.navy.mil/HomePage/GRAB/GRAB.html.

12. See Naval Research Laboratory, *Galactic Radiation and Background (GRAB) Satellite Declassified,* NRL Press Release 41-98r, June 17, 1998; Keith Hall, prepared remarks at the Naval Research Laboratory 75th Anniversary Event, June 17, 1998. Also see Bruce Berkowitz, "The Nine Lives of Slick Six," *Air & Space/ Smithsonian,* February–March 1997.

13. Kevin C. Ruffner, ed., *CORONA: America's First Satellite Program,* Washington, D.C.: Central Intelligence Agency, 1995, p. xiii. Ruffner, a member of the CIA history staff, headed the effort to assemble a declassified history of CORONA. When he did his research for the project, he had access to the National Reconnaissance Office and National Photographic Interpretation Center files, and was cleared for NRO systems. However, he had never heard of GRAB (a program that had ended twenty years earlier) and did not have any reason to search for records about the program.

14. Kenneth E. Greer, "CORONA," *Studies in Intelligence,* Talent/Keyhole classified supplement, Spring 1973, pp. 1–37. Reprinted in declassified form in Ruffner, *op cit.*

15. See Ruffner, p. 1.

16. See the Woods Hole Web page at www.whoi.edu/home/.

17. See Roger C. Dunham, *Spy Sub: A Top Secret Mission to the Bottom of the Pacific,* Annapolis, MD: Naval Institute Press, 1996. Dunham, currently a physician and author, was a crewman on the *Halibut* and wrote his memoirs, which the Defense Department cleared for publication. The Defense Department had Dunham identify the submarine as the fictional *Viperfish,* but Dunham describes the boat as a nuclear-powered submarine that once launched Regulus cruise missiles, and there was only one ship that fits the description. Also, Dunham includes a picture of the *"Viperfish."* It's the *Halibut.* Also see Sherry Sontag and Christopher Drew, with Annette Lawrence Drew, *Blind Man's Bluff: The Untold Story of American Submarine Espionage,* New York: Public Affairs, 1998, which deals with the *Halibut* and its "fish" at length.

18. Harold F. Edgerton, "Edwin Albert Link," in *Memorial Tributes: National Academy of Engineering, Vol. 2,* Washington, D.C.: National Academy Press, 1984, pp. 173–178.

19. See U.S. Coast Guard Marine Board of Investigation Report, *Entanglement of the Submersible Johnson Sea Link with Submerged Wreckage off Key West, Florida, on or About 17 June 1973 With Loss of Life,* Report No. USCG/NTSB-Mar-75-2, March 12, 1975.

20. See Robert D. Ballard with Will Hively, *The Eternal Darkness,* Princeton, N.J.: Princeton University Press, 2000, pp. 218–254; and Robert D. Ballard with Malcolm McConnell, *Explorations: A Life of Underwater Adventure,* New York: Hyperion, 1995, pp. 209–236.
21. Quotations are from interview with Robert Ballard, August 1, 2002.
22. William J. Broad, "Titanic Wreck Was Surprise Yield of Underwater Tests for Military," *New York Times,* September 8, 1985.
23. Several accounts of JENNIFER have been published, including Roy D. Varner, *Matter of Risk: The Incredible Inside Story of the CIA's Hughes Glomar Explorer Mission to Raise a Russian Submarine,* New York: Random House, 1979, and Clyde W. Burleson, *The Jennifer Project,* College Station, TX: Texas A&M University Press, 1997. William Colby discussed the project and how it was revealed in his biography with Peter Forbath, *Honorable Men: My Life in the CIA,* New York: Simon & Schuster, 1978. The CIA confirmed the mission, but not the details, when Director of Central Intelligence Robert Gates gave President Boris Yeltsin a film of the CIA interring the remains of Soviet sailors recovered in the operation at sea.

INDEX

U.S.S. Halibut, 214
U.S.S. Scorpion, 213, 215, 217
U.S.S. Tecumseh, 79
U.S.S. Thresher, 212, 215, 217
UNIX operating system, 157

van Cleve, William, 37
Vandenberg Air Force Base, 211
Vergeltingswaffen-2 (V-2), 86, 87
Versailles Treaty, 86
Victoria, Queen of England, 179
Vietnam War, 34, 38–40, 42, 76, 88, 120,
    151–152, 200
Viking rocket, 208
Virtual exploration, 215
Viruses and worms, 143–147, 156–157,
    160–161, 163, 176
Von Braun, Wernher, 86, 87
Von Trapp, Georg, 82n

Walker, John, 185–186
Wall Street Journal, The, 9, 59
Wal-Mart, 111–112
Wang Wei, 156, 158
Warfighting (Marine Corps), 107
WarGames (movie), 135–136
"War Next Time, The" (Rona), 59
War of the Worlds, The (Wells), 155
Washington Post, The, 203
Wayne, John, 108
al-Wazir, Khalil, 128
Weapons of mass destruction, 17, 127
Weapons Systems and Information War
    (Rona), 30
Weather Bureau, 54–55
Weiland, Ric, 29
Wells, H.G., 155
Weyrich, Paul, 60, 106n
Whitehead, Agathe, 82n
Whitehead, Robert, 81–83, 87
White House Office of Science and
    Technology Policy, 37
Wiesel, Elie, 23
Wigner, Eugene, 23
Wigwag signaling system, 52–55
Williams, Margot, 203–206
Wilson, Peter, 138–140, 155
Wilson, Woodrow, 184

Winding Down, 105–106
Windows (operating system), 144
Winslett, Kate, 212
Wired, 146, 156
Wohlstetter, Albert, 33–34, 68, 75
Woods Hole Oceanographic Institute, 213,
    215, 216, 217n
Worldcom, 31, 177, 220
World Factbook (Central Intelligence
    Agency), 4
World Trade Center, New York
    1993 bombing, 11, 13, 18
    September 11, 2001, 8, 17–18
World Trade Organization, 125
World War I, 182
    electronic warfare in, 184–185
    trench warfare in, 76, 183
    Zimmerman telegram and, 183–184, 196
World War II, 24, 31, 32, 103, 129
    Battle of Britain, 25, 86
    electronic warfare in, 25–27, 36, 185
    glide bombs in, 120
    Pearl Harbor, 17, 153, 155
World Wide Web. *See* Internet
Wright, Bruce, 150
Wright, Orville, 214
Wright brothers, 84
WS-117L, 210, 211
WYSIWYG, 159

Xerox Corporation, 62, 159

Yamamoto, Isoroku, 129
Yemen, 18, 99, 100–101, 117, 123, 130
Yesler, Henry, 27
Yom Kippur War, 197
Yoran, Elan, 171
Yousef, Ramzi Ahmed, 13
Yugoslavia, 3

Zama, Battle of, 19
Zapatistas, 74
Zapping (precision strike), 76–77, 95,
    97–99
al-Zawahri, Ayman, 197–199
Zhawar Kili, Afghanistan, 122
Zimmerman, Arthur, 183–185
Zimmerman, Philip, 187–188

ABOUT THE AUTHOR

Bruce Berkowitz is a research fellow at the Hoover Institution at Stanford University and a senior analyst at the RAND Corporation. He began his career at the Central Intelligence Agency and served as a staff member for the Senate Select Committee on Intelligence. Since leaving government, Berkowitz has combined careers as a widely published author and a consultant to the Defense Department and other government agencies.

Berkowitz is a frequent contributor to *The Wall Street Journal* and has published articles in *Foreign Affairs, The National Interest, Foreign Policy, Technology Review,* and *Issues in Science and Technology,* the policy journal of the National Academies of Science and Engineering. He also writes regularly for the *International Journal of Intelligence and Counterintelligence,* where he is a member of the editorial board, and *Orbis,* where he is a contributing editor.

Berkowitz received his bachelor's degree from Stetson University, graduating summa cum laude, and earned his masters and doctorate at the University of Rochester. His home is in Alexandria, Virginia.